Human Wildlife

The Life

Human Wildlife

That Lives On Us

Dr. Robert Buckman

First published in English in Canada by Key Porter Books Limited, Toronto, Canada, 2002
 Published 2003
Printed in Canada on acid-free paper
9 8 7 6 5 4 3 2 1

The Johns Hopkins University Press
2715 North Charles Street
Baltimore, Maryland 21218-4363
www.press.jhu.edu

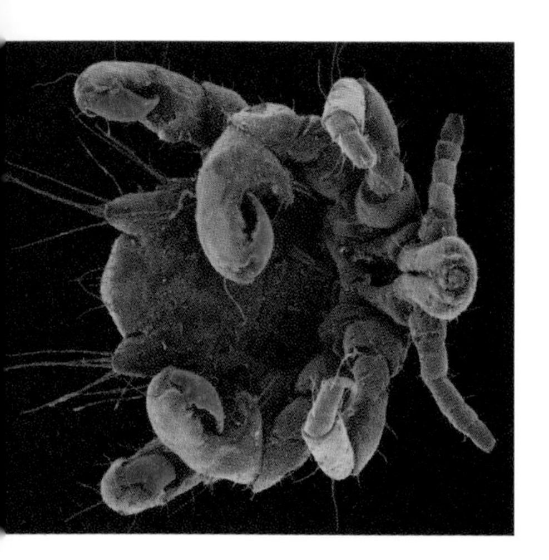

ISBN 0-8018-7406-8
ISBN 0-8018-7407-6 (pbk.)

Library of Congress Control Number: 2002113826

A catalog record for this book is available from the British Library.

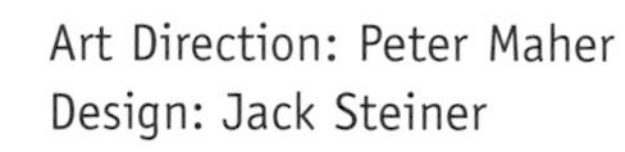

Art Direction: Peter Maher
Design: Jack Steiner

Table of Contents

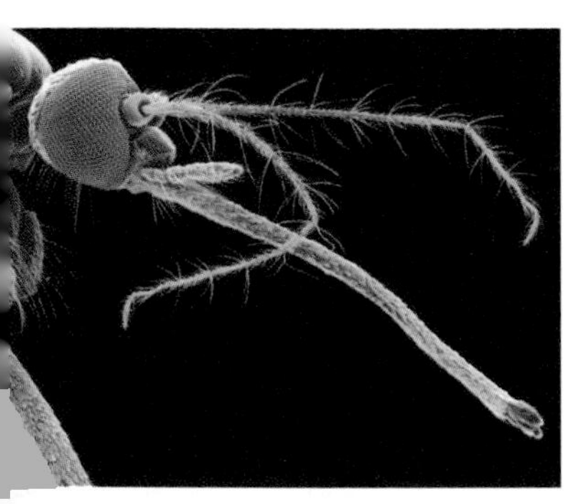

1 Planet Human

We are not alone in our skin.

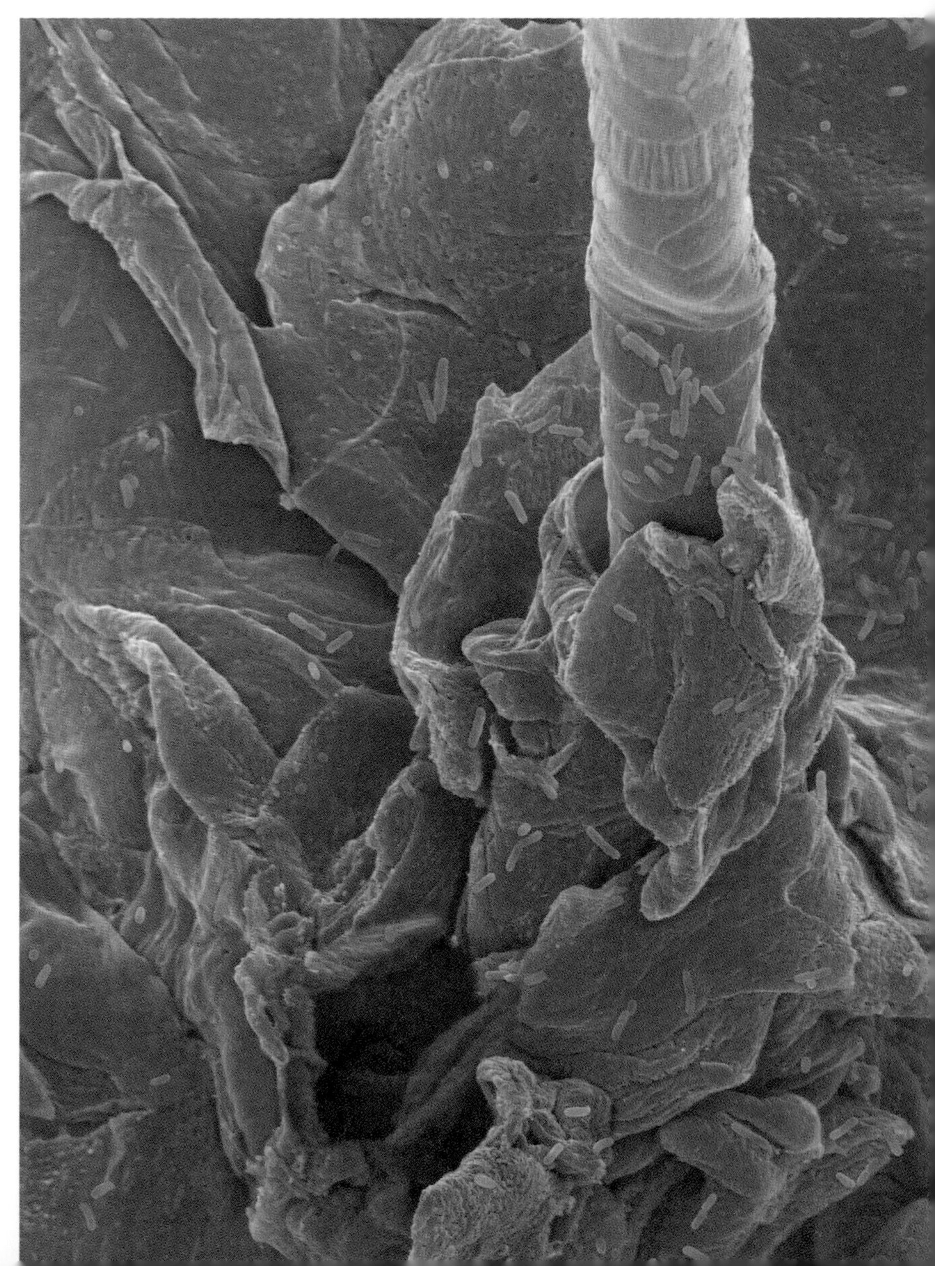

Skin and some of its residents. A human hair follicle and some bacteria harmlessly loitering on the surrounding skin.

We human beings are a part of the life on this planet—not apart from it. Not only do we live among and between other life forms, but many life forms live in, among, on, up, and inside us. In fact, from the viewpoint of those life forms, a human body is a planet in itself. In many respects, then, this book is an atlas—albeit a somewhat idiosyncratic and erratic one—of Planet Human and of some of the life forms that inhabit it.

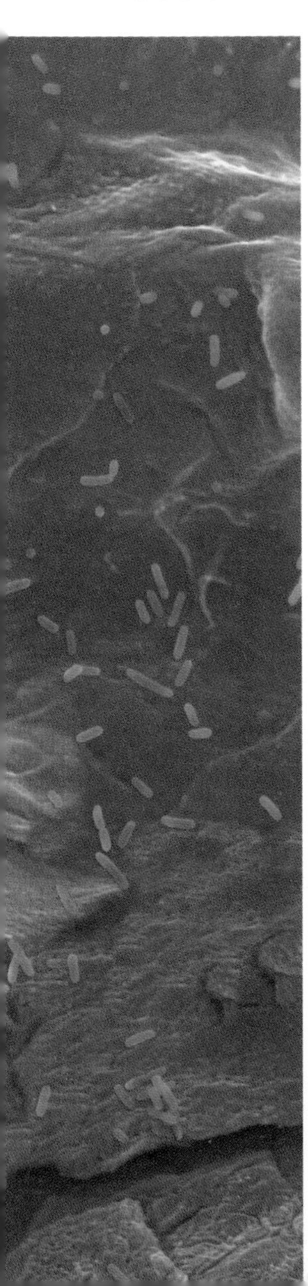

Our planet has vast numbers of harmonious ecosystems, specialized environments in which multiple species coexist without interfering with (or destroying) one another.

The number of living species on this planet has been estimated (by the United Nations Global Biodiversity Assessment) as somewhere in the range of 10 to 100 million. And lots of them live together—some in harmony, some not.

So, let's get basic about the way in which different species cohabit and share territory—whether in a tropical rain forest or in a human armpit. The correct scientific word for two or more species living together is *symbiosis,* and there are different categories of it. In one type of symbiosis, two or more species just occupy the same environment and may even benefit each other to some degree. This is called *mutualism*. The little birds that perch on the hippos' back and eat the mucky little worms and bio-garbage they find there are examples of mutualists. In another type of relationship, one species is totally dependent on the other but does no harm to its host—just as in medieval times, a poor person might subsist on crumbs from a rich man's table. This type of relationship is called *commensalism* —the eyebrow mite *Demodex* is a good example (as we

Peace and Quiet. Our planet has vast numbers of harmonious ecosystems, specialized environments in which multiple species coexist without interfering with (or destroying) one another.

shall see in the next chapter) in that it apparently does exactly nothing except sit there.

Then there is *parasitism*—in which one species depends totally on the host, to the detriment of the host—be it major damage or minor inconvenience. The best kind of parasite (in terms of biological success, that is) produces an effect on the host that is so minor it is scarcely noticeable—for example, most of the common external parasites that live on birds (in their feathers and in their nests). Other parasites cause very little damage most of the time, but can occasionally be a problem—for us humans, the roundworm *Ascaris* is an illustration of this; as testament to its success, it is estimated that *Ascaris* is present inside about one and a quarter billion human beings at this moment.

Then there are some parasites that are so nasty and so dangerous that they kill the host even before the poor host has had a chance to pass the parasite on to new victims. The Ebola virus is a good example of this very unsuccessful (from the biological point of view) parasitism that often happens when a parasite has only recently moved in on a new species of victim.

This is the range of the relationships that we shall be talking about—from the eyebrow mite to the Ebola virus, from *Ascaris* to malaria, from *Streptococcus mutans* to the guinea worm. They are all residents of or visitors on, in, or around our bodies—and the more we know about them, the more insight we gain about ourselves.

Parasitism—A Wildly Popular Lifestyle

A reassuring note: Please don't be threatened by all this talk of parasitism. It is a very common form of relationship between species. Of the 10 to 100 million species that exist on earth, at least three million species are parasites—that is, they gain their sustenance in one way or another from their host. To coin a variation on the ancient physicians' phrase, when it comes to parasitism, "there's a lot of it about." Even if you think of humans as parasites on Planet Earth (a rather unkind and reductionist view, the way some people express it), we are still in good company—there another 2,999,999 other species doing it, too. This isn't any excuse, but at least we're not the only ones.

Our Body Politic

If the numbers of species on Planet Earth are staggering, the numbers on or in Planet Human are hardly less so. For example, the body space of an average adult human being comprises approximately 100 trillion cells—that is one hundred million million separate units of living matter. This is a fairly impressive number. Even more impressive, however, is the fact that of those 100 trillion cells inside the average human frame, only 10 trillion are human cells. The other 90 trillion cells are bacteria (with a few other parasites, fungi, and miscellaneous riff-raff thrown in for good measure). Inside our own bodies we are outnumbered by other species nine to one. Fortunately, the human body is not a democracy, so even though our bodily bacteria do influence our workings in many ways, they don't have a vote. They therefore cannot decide—on their own—to throw us out entirely (although on occasions they can cause a variety of expulsions and upsets and ultimately, if one cares to think of it that way, they can cause revolution, dissolution, and redistribution).

Yet, even accepting that some species have the potential for doing us considerable harm, we can perhaps afford to be a little fairer to many of the other less threatening species with whom we share our body (and, in some cases, our planet). Not everything that is non-human is necessarily bad for us. The mood of recent times has been to regard every non-human species in or on our bodies as untrustworthy and threatening. This is undoubtedly true of some species: there is no such thing as a friendly smallpox virus, and you cannot domesticate a malarial parasite and have it come when you call it. Nevertheless, we have tended to take the attitude that all other species exist on this planet solely to be eaten, looked at, or cuddled, and if they are not suitable for any of the above, then we should be rid of them. Without getting too polemic about this, one could call this tendency to regard other species as inherently dangerous as a sort of "eco-political speciesism" and perhaps now is a good time to evaluate the foundations for that attitude. As will become clear during the course of this book, we need to know more about other species so that

A Highly Unexpected Statistic

The number of different species on this planet is estimated to be between 10 and 100 million: of those, more than 3 million are parasites.

we can distinguish the ones that that we really do need to avoid from those that—however strange or alien they may appear, and however much we would not like to have them as pets or as food—are perfectly satisfactory as co-tenants of our ecosystem.

Now I don't want the pendulum to swing too far the other way. While I am all in favour of decency and fairness, there are limits. I don't want to be seen as some knee-jerk trans-species appeaser and wuss. I still believe that if any of us happen to have a bothersome tapeworm inside us, we should be able to take the appropriate medication straight away and not have to get the tapeworm's opinion first. While I freely acknowledge that worms were on this planet long before the human species (see Chapter 4), I also acknowledge that as regards my own bowel, I owned it first. Any visitors to it are to be treated as newcomers and must not create a major disturbance or I will exercise my owner's rights.

Which brings us on to the most important point and the central focus of this book: what constitutes a genuine disturbance? Under what circumstances should we choose to exterminate the invader, and when is it reasonable to pursue the path of peaceful coexistence? If, for example, I pick up a tribe of cholera bacteria and they proceed to multiply in my bowel and threaten my life, the word "disturbance" would obviously be an appropriate descriptor. No sane individual would advise me in the middle of a bout of cholera to simply "be nice and try to co-exist with these strangers." The motto "live and let live" is simply inapplicable, mainly because the cholera bacilli (albeit unintentionally) reduce the chances of their human host living. At one end of the spectrum, therefore, it is easy to see that the collision of the two worlds—human and wildlife—constitutes a disturbance.

At the other end of the spectrum, there are a whole smorgasbord of wildlife species—fungi, saprophytic bacteria, and dozens of others—living and flourishing on my skin that do not constitute a disturbance. Commensals are not vandals, and we should not try to exterminate them.

In between the two extremes is the entire gamut of wildlife that lives on, in, or around us: the local inhabitants and residents, the itinerant visitors, the longer-term guests, the squatters, the mildly irritating vandals who spray paint slogans on the wall, and the occasional invading army of heavily armed and ruthless troops who regard us as fertile territory for colonization and exploitation. The important thing is to work out which is which, and to distinguish between harmless friends and lethal foes. This book will help you to do that.

By the time you've read this book—and looked at the highly unusual pictures—you will be able to recognize some of the features that distinguish the behaviour of the docile residents from the invading mercenaries. The key to the whole thing is knowing what you're dealing with, how it behaves, and what it might—or might not—do. At the end of our brief reconnoitering trip around Planet Human, you'll have seen at least some of its denizens and had a glimpse of the complex relationships between human wildlife and human life.

The Grand Scale of Things

What you see here is an approximate scale showing the sizes of some of the bodies that we shall be talking about in the book. Personally, I find that I need a visual scale to help focus my mind. It is helpful to know how big a bacterium is compared to a human red blood cell, or how big a mite is compared to the width of a human hair.

Of course, there are some comparisons that are now so clichéd that they no longer hold any meaning at all. (For example, we always used to hear of statistics that compared the surface area of a person's lungs to the size of the average tennis court. It always made me think that if the average human's lungs were opened and spread over a tennis court, it would drastically affect his breathing. So you won't find any tennis courts on this scale.) In any event, I hope this scale is useful as a reference as we meet the various organisms.

Size is everything. A few common objects and their relative sizes.

Bacteria on a pinhead ×400

Bacteria on a pinhead ×500

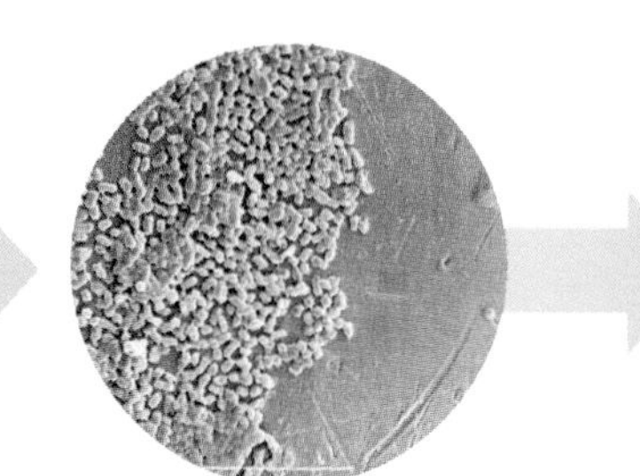

Bacteria on a pinhead ×1000

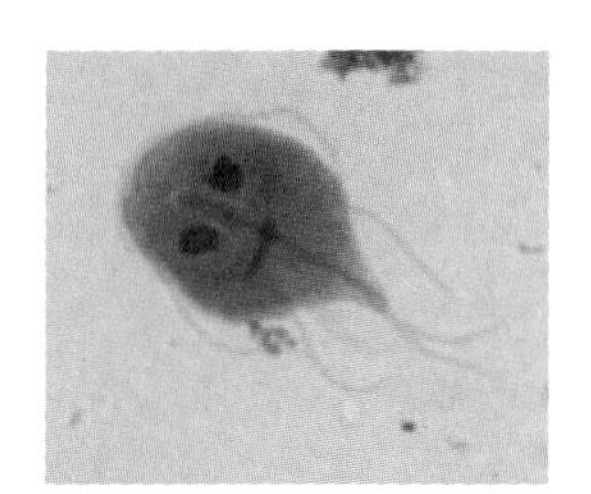

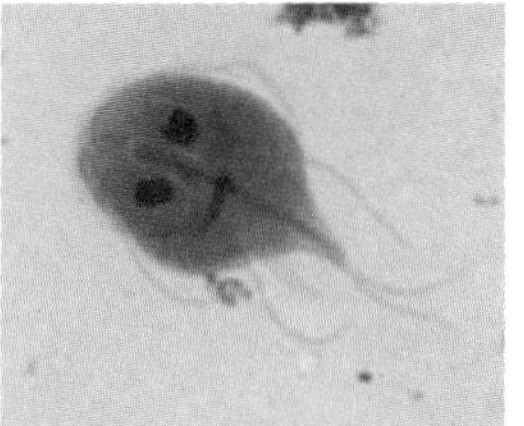

Giardia lamblia

114 *Giardias,* side-by-side, could comfortably swim through the eye of this needle without creating a protozoan traffic jam.

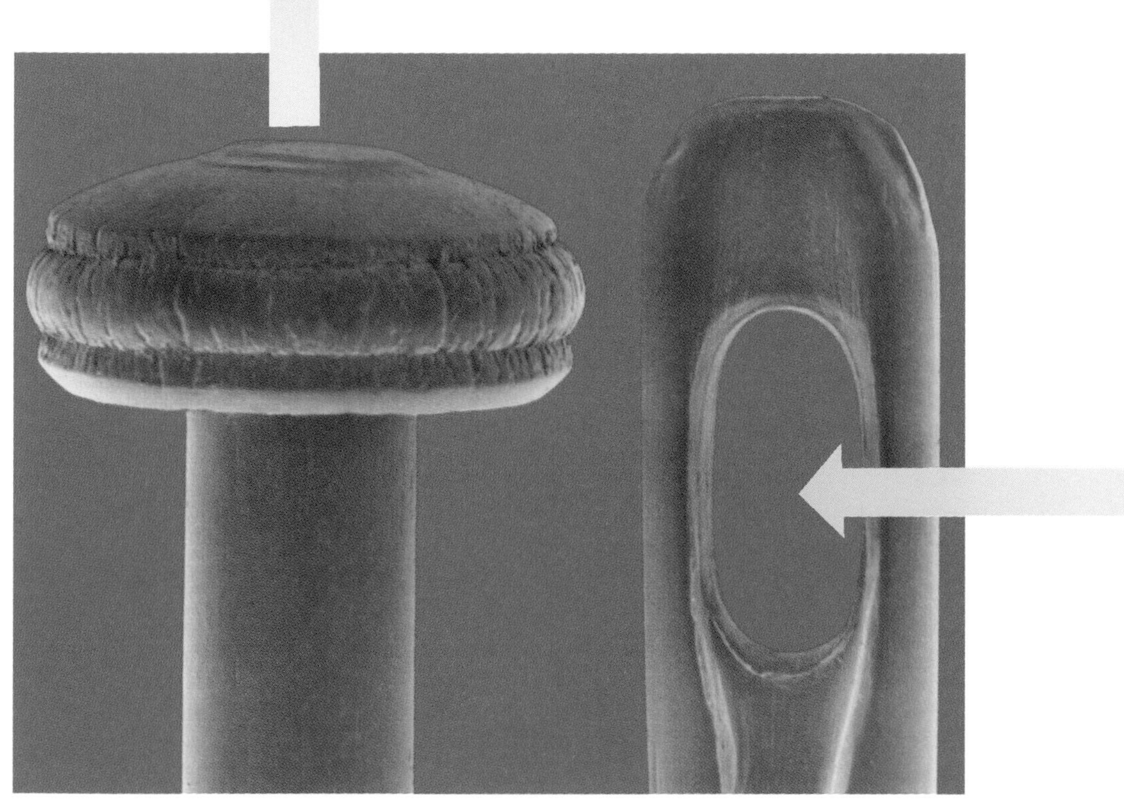

Pinhead and eye of a needle

2 The Meaning of Lice

One of nature's longest running soaps is the one that features (in no particular order) mites, ticks, fleas, and lice—it's called "The Small and the Restless."*

*The ratings are very high.

Head Louse. This animal *Pediculus humanus capitis* is extremely good at holding on firmly to the thin hairs on the human head. It's so good at it that it stays there for most of its life.

It's a Smallish World After All

Most of the animals you are about to see are only just visible to the naked eye (as if there were such a thing as an "a clothed eye" or "a fully-dressed eye"), yet their importance is not related to their size. While most of them do very little harm, a few are capable of carrying diseases that can be quite serious. Hence it is important to be aware of the more dangerous types of parasites, and under what circumstances you should take particular precautions. Parasites are very adept at evading our responses to them—picking them off when we see them, for example; their small size and defensive mechanisms make our job more difficult, and theirs easier.

Mite Is Right

The word "mite" is derived (probably) from an Old English word meaning "very small and basically worthless." It was applied to coins of very small denomination in the fourteenth century—when small denominations were

MITE BITES Various members of the mite tribe. The dust mite (left) and *Demodex,* the eyebrow mite.

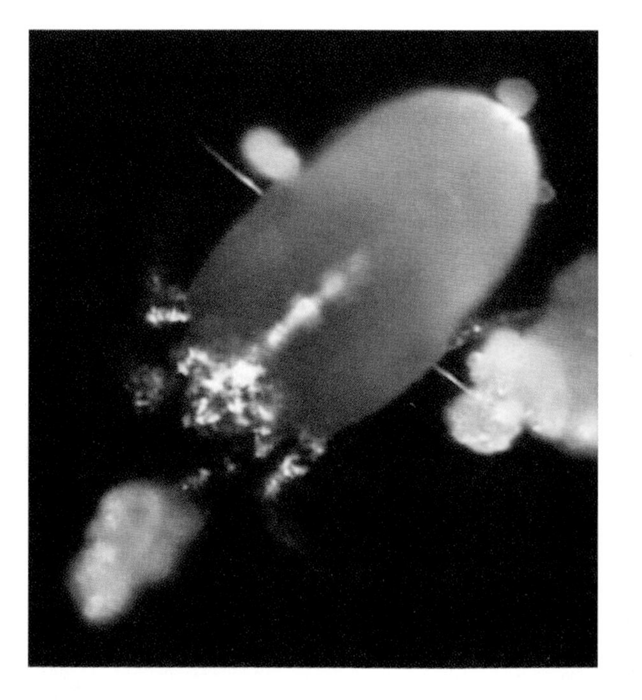

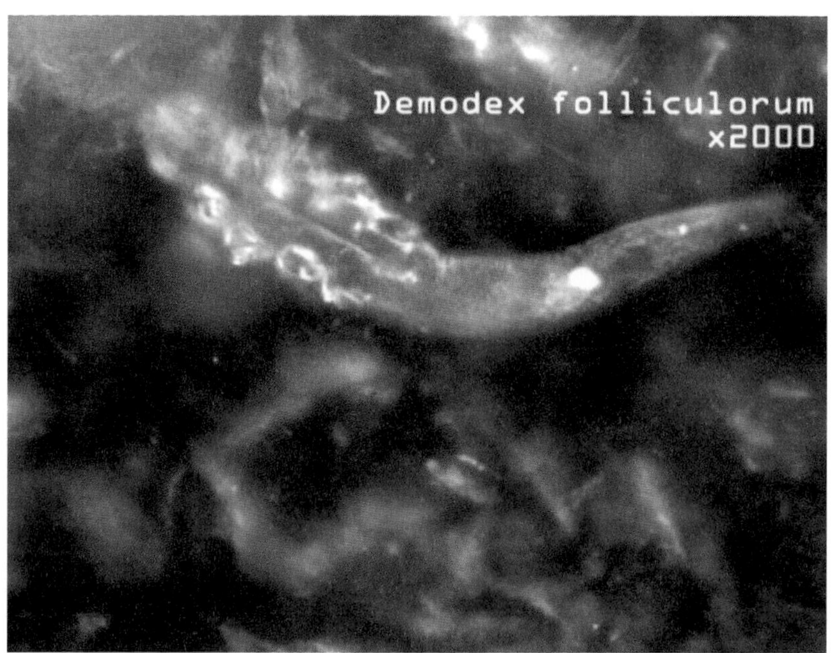

the only ones around—and even today it can mean "a pittance," as in a "widow's mite." On the other hand it can also mean any arthropod belonging to the *Acarina* family, a group partly related to spiders. A few representatives of the mite family have been kind enough to pose for the portraits on this page—and we shall visit three of them in some detail.

Eyebrows Will Be Raised!

Of all the creatures that our team encountered during the preparation of this book, perhaps the most unlikely was this tiny little cigar-on-legs that happens to live in your eyebrows. By the time we're in late adulthood more than two-thirds of us will have a few of these creatures snugly burrowed into our eyebrow or eyelash follicles. Its name is *Demodex* and as you look at its photograph you need to know that the legs are—

On Its Way Home. This cigar-on-legs is an eyebrow mite, *Demodex*, and is probably on its way home to an eyebrow follicle after a little outing.

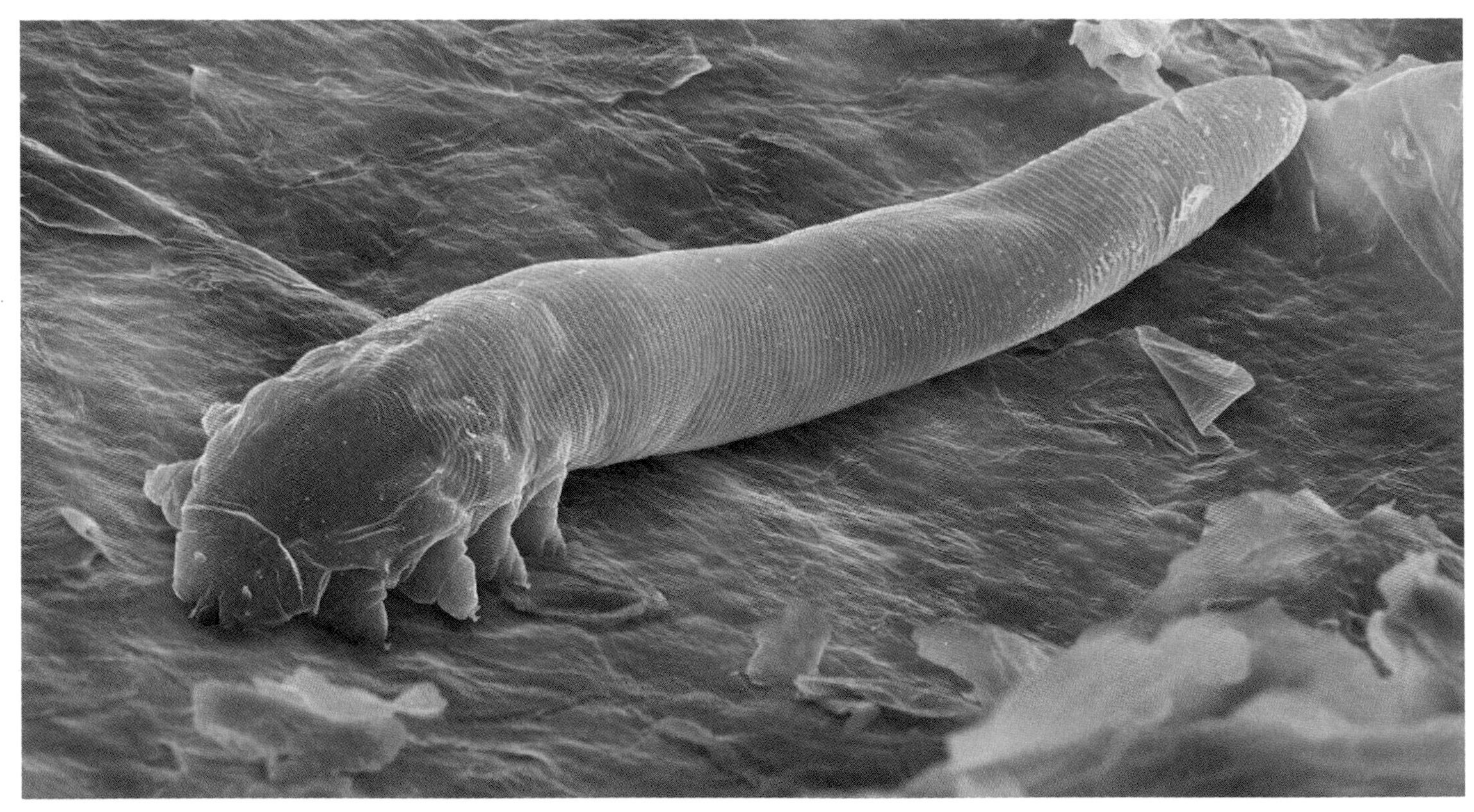

unexpectedly—at the front end. It waddles along fairly slowly, presumably dragging its sorry butt along behind it, until it comes to the follicle of one of your eyebrow hairs or an eyelash. It then burrows down head-first into the follicle, perhaps accompanied by the mite equivalent of puffing and panting, followed by an inaudible sigh of welcome relief. Its wriggling and squeezing entry does almost no damage, and it just squats there eating the skin cells that are the biological equivalent of trash it finds lying around, and doing nobody any real harm.

Hear No Evil, See No Evil, and Try to Do No Evil

As the mite burrows in it leaves its bum sticking up in the air, as you can see from the photo on the next page of three *Demodex* mites (their faces are hidden inside the follicle so we can't give you their names even if we could tell which was which—which we can't. And neither can they, so that's all right.) They are genuinely harmless and we call that kind of symbiosis in which one party gets nourishment and shelter without harming the other party commensalism. So *Demodex* are typical commensals, as we mentioned in Chapter 1—they're there, but they're not doing any noticeable damage. They are the mendicants who make a meal of crumbs from the rich man's table, the snappers-up of unconsidered trifles.

Or so we believe to be the case most of the time. It is possible, though by no means certain, that if you happen to have lots of *Demodex* mites and you are a heavy user of eye make-up, some of the *Demodex* may carry flakes

The Face In Your Face. This is *Demodex,* smiling for the camera.

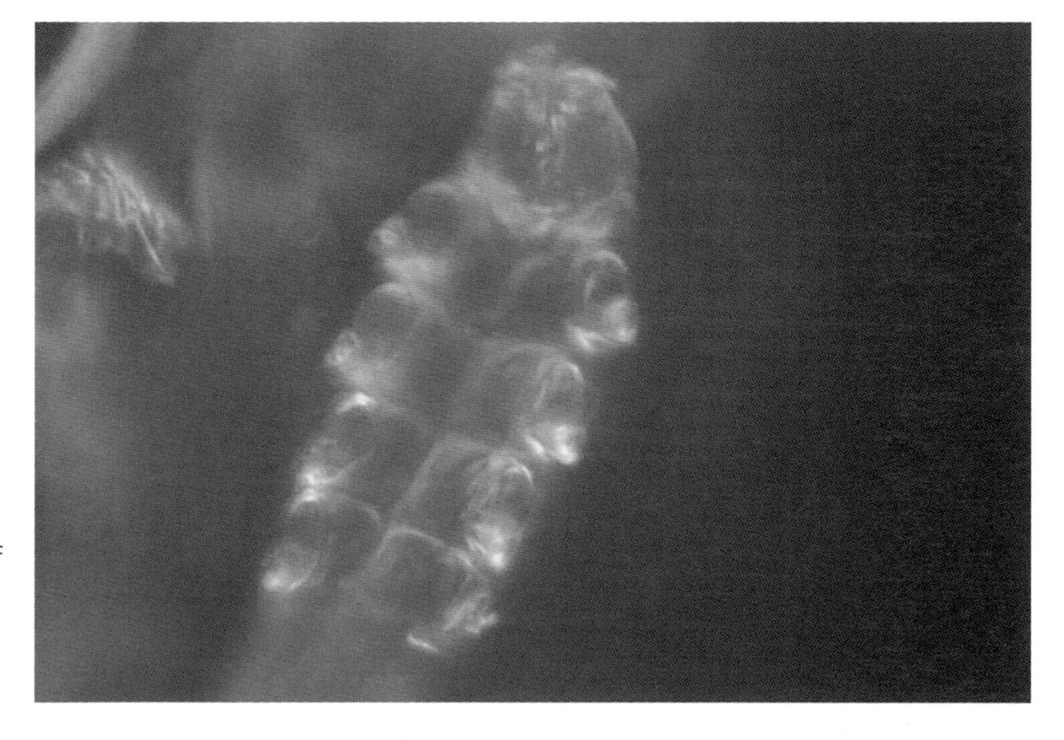

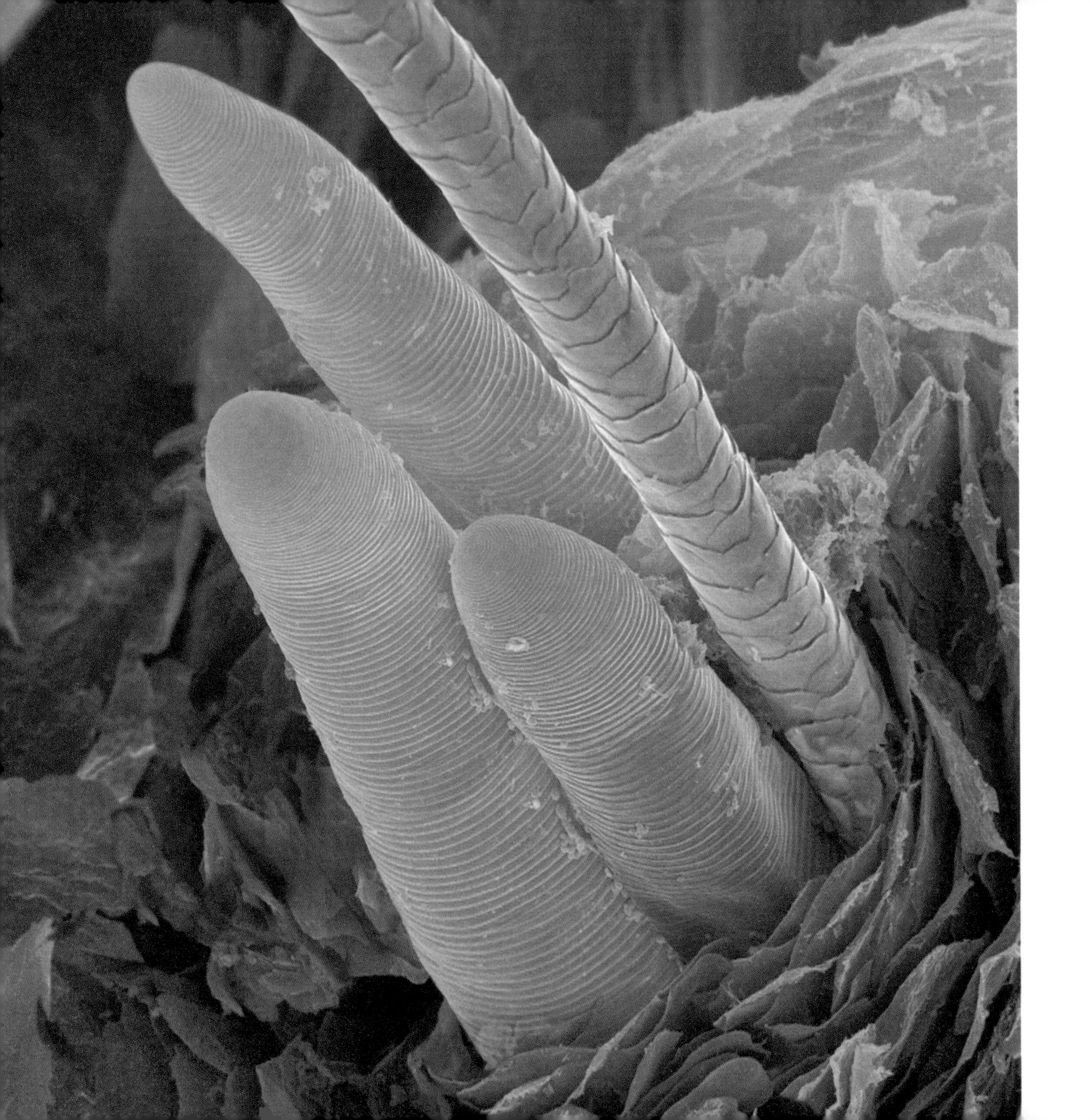

of your mascara or whatever deeper into your follicles, which may set up a mild inflammation. If this happens, it is probably rare. No one has ever seen *Demodex* mites at the site of such a problem, so no one actually knows what a *Demodex* mite looks like when it is wearing a lot of mascara. We can only imagine.

(Opposite)

Larry, Curly, and Moe. Three *Demodex* mites heads-down in a hair follicle. The one farthest from the camera is probably Curly.

What Could Get Up Your Nose? The House Dust Mite Might

Let me frame for you one of life's most unfathomable mysteries, an eternal question that—until recently anyway—seemed to escape every line of human inquiry: Why is the dust in everybody's house the same colour? Why is the dust the same grey no matter whether the carpet is red or green, and the curtains yellow or blue and the sofa green or brown?

The answer may come as a somewhat sobering thought: it is because house dust consists almost entirely of one thing—your used-up and dead skin. While you walk around your house, even without doing anything vigorous in the way of scratching or rubbing against the furniture, you are shedding thousands of worn-out dead skin cells from the top layer of your skin into the air. The dead cells are very thin and almost transparent and when you have millions (or billions) of them, they look grey, almost silvery. Which is what gives that characteristic colour to the contents of your vacuum cleaner. It is a constant reminder of the frailty and mortality of all human biological components that what you empty from your vacuum cleaner today was covering your leg yesterday.

This piece of information might be nothing more than an irrelevant piece of trivia were it not for the fact that a variety of creatures call house dust their home, and the commonest inhabitant, a mite, has an unfortunate knack of causing illness in certain people.

The House Dust Mite living comfortably on our dead skin cells. The mites pictured here were happily moving in between the fibres of a cotton sheet (the background) looking for more food.

The creature shown in the photo at right is the common house dust mite, *Pteronyssimus dermatophagoides*. Its name is long, but the creature itself is extremely short. In fact the average adult is about 0.4 of a millimetre—if it stood up it would be about ten hair-breadths high. There is

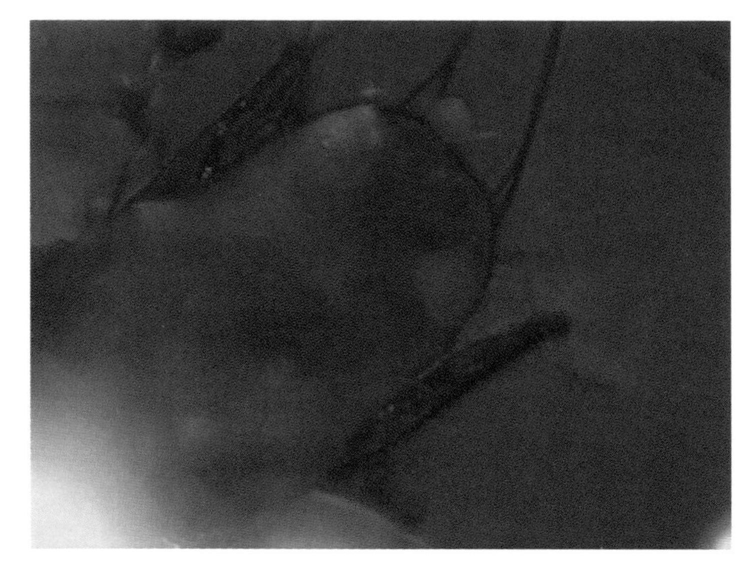

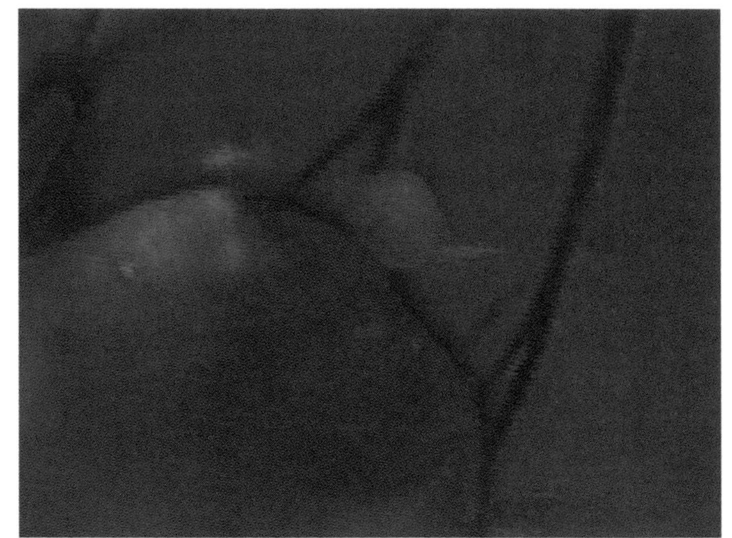

A Sight Rarely Seen. The actual cause of asthma is the fecal pellet of *Pteronyssimus,* which is the object that looks a bit like a light-bulb dangling from the posterior.

one thing about the house dust mite that causes a lot of us real trouble. To be blunt about it, it's their poo. The front end of the mite presents no major problems: they don't bite or sting or inject venom. They can't paralyze us, give us Ebola virus, or put us into septic shock, but their feces can give some of us asthma.

The accompanying photo shows something that (as far as we can tell) has never been by the public before: a house dust mite in the process of defecating. A *Pteronyssimus* at the toilet (or should that be Ptoilet?). Curiously it is the fecal pellet—the little light-bulb-like blob at the back end of the animal—that is the source of the problem. About one-third of the human population (and this proportion varies from survey to survey) are allergic to components of house-dust-mite feces. This is established by a skin-prick test in which an extract of ground-up mites is put under the skin with a small, and virtually painless, needle. If the person gets a red bump on the skin measuring about 4mm, then that person is allergic to house dust mites.

The difficulty is that the mite has evolved a rather cunning way of preparing its food that relies on a digestive enzyme (*der-p1*) in its feces that starts the breakdown of skin cells. In other words, the house dust mite marinates its food source in its own feces. This may not be an optimal system of food preparation from our viewpoint—it's hardly the etymological equivalent of Martha Stewart and the Cuisinart—but it works for them. Unfortunately, however, it does us some damage.

As Dr. Adrian Custovic in Manchester, England, clearly and enthusiastically explained to me—the digestive enzyme that assists the house dust mire in its preprandial meal preparation is really quite harmful to the lining of our bronchial tubes. It loosens up the junctions between the cells lining the airways and allows the components of the mite feces (as well as other materials) to penetrate deeper into the walls of the airways. These materials are now in a better position to trigger an asthmatic attack, which, as we all know, can in some cases be quite severe. Being aware of the source of the problem can actually help you and if you or a member of your family has asthma, then it's worth looking at the box on this page for some practical tips.

Practical Points About Asthma

1. Do you (or does a family member) have asthma? Take the advice of your physician as to whether you have the condition and what you should be doing about it.
2. Are you allergic to house dust mite extract? (This can be established by a skin prick test.)
3. IF you are asthmatic AND you are allergic to house dust mites, then it's worth taking the appropriate precautions (e.g. buying special close-weave mattress covers and pillow covers, considering taking up the carpet in your bedroom and using hardwood floors, vacuuming often, and so on).
4. ALWAYS, ALWAYS take your doctor's recommendations seriously about your asthma and use the medications REGULARLY as prescribed.
5. In the event of an asthma attack, seek medical advice EARLY if the attack is not responding to your usual medications.

Scabies: The Little Mite that Could—and Does

About three hundred million human beings harbour this nasty little mite that causes the very itchy skin condition we call *scabies.* This mite is very small—and very peculiar. Tinier than the head of a pin, the female gets under your skin—in the fold between your fingers or toes, for example—and burrows steadily along under the skin on a one-way journey that will

Neither a Burrower. . . The female scabies mite takes no notice of any advice and burrows under your skin wherever she can. Too deep to be harmed by your scratching, she lays eggs along the track of her burrow and then dies. Her children pick up the flag and crawl back up the track to infest someone else.

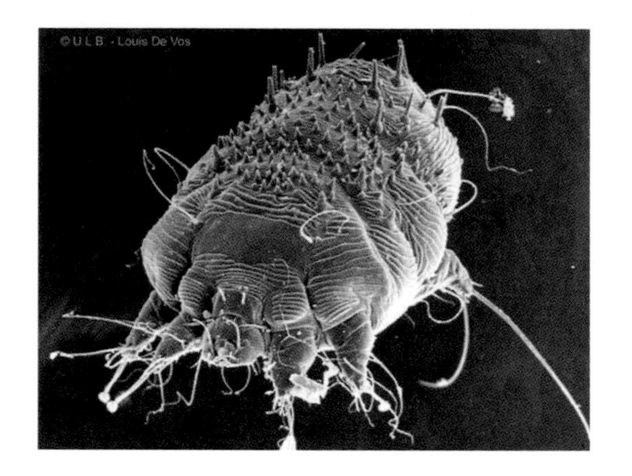

lead ultimately to her death. But don't waste any sympathy on her because as she goes along on her journey she lays eggs to start the next generation. After some days, the eggs hatch and the even tinier infant scabies mites start the reverse journey—burrowing back to the surface of your skin, where they will transfer to another victim, then burrow under the new landlord's skin, lay eggs, and die just as their mother did before them.

It is the track that the adult female makes on her journey that causes the human victim intense itching—a symptom is even worse if you have had a previous case of scabies and produce a greater inflammatory reaction to it.

The scabies mite—*Sarcoptes scabei*—is so small that it can only just be seen by the naked eye. It has eight legs, although how something that small can coordinate eight legs without getting confused or tripping over itself is not precisely understood (by me, anyway). It is passed from human to human only by prolonged contact; a handshake or a hug will not do it (though it can do many other things). In fact, during World War I, experiments were done in which Conscientious Objectors were ordered to spend the night in a military cot with a soldier who had scabies (things were different in those days, you know.) The Conscientious Objectors developed scabies, proving that skin-to-skin contact lasting several hours was required for transmission.

The old song "I've Got You Under My Skin" is still popular and well-loved, but the scabies mite is not.

Deer, Oh, Deer

The mites that we have discussed in this section so far belong to the same group of arthropods as ticks—although perhaps most of us have a murkier and less clear image of what a tick actually is. By and large, ticks are a more serious threat to us, mainly because of the diseases they carry.

A good example of the role that ticks play in human illness and misery can be seen by looking at the black-legged deer tick. It's the tick that has the black legs (although they're really a sort of smudgy dark brown, in my opinion), not the deer (the deer has light brown legs and very fetching they look, too). Although a black-legged tick that lives on brown-legged deer doesn't sound as if it has anything to do with humankind, in fact it does, and it's serious. The deer tick just happens to carry a nasty bacterium called *Borrelia burgdorferi*, which is the corkscrew-shaped agent that causes the quite nasty Lyme disease (named after the town in New England where it was first described). The story is an interesting one—with deer, mice, and acorns in starring roles.

Ticked Off. Deer ticks are potentially dangerous. They are not a problem for deer, but they are a hazard for humans because they carry the bacteria that cause Lyme disease.

As Dr. Rick Ostfeld, an ecologist with the Institute of Ecosystem Studies in Millbrook, New York, points out, there is a fascinating cycle of life and disease within the forests of New England (and other areas). When the oak trees give a plentiful yield of acorns, the number of cases of human beings getting Lyme disease goes up in proportion. In years

Not a Hollywood Creation. Under the microscope, the tick looks like something out of *Star Wars*. Unfortunately ticks are not limited to fantasy; the problems they cause are real.

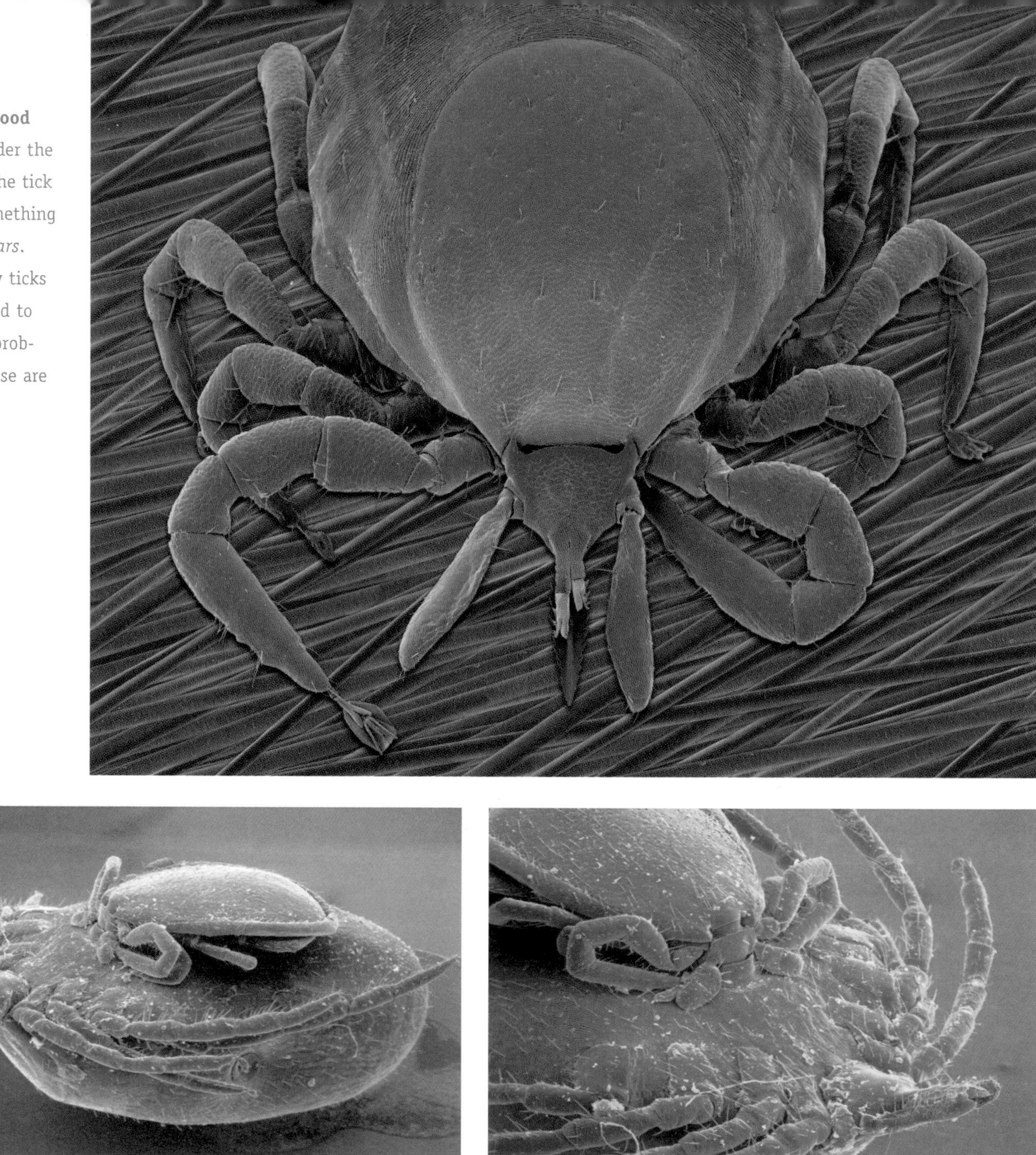

Missionary Tricks. We do know that the male is the small one on top and the female is the larger one underneath.

when the oak trees are mean and stingy about their acorn yield, there are fewer cases of Lyme disease in humans. Why?

The answer lies in the unshakeable relationship of supply and demand. It so happens that acorns are a favourite food of the mice that freely run around on the forest floor. When the yield of acorns is high, the mice have a plentiful food supply and they flourish and reproduce (after all there is very little else for them to do in the way of entertainment since they don't have cable TV or bars). A large and plentiful supply of mice in the forest suits the black-legged deer tick extremely well. This would not be of human concern except for two facts: first, the field mice happen to carry the bacterium that causes Lyme disease in humans, and second, humans inadvertently step into the life cycle of the tick by walking in the woods.

Dr. Rick Ostfeld

The *Borrelia burgdorferi* bacterium, which the field mice in the forests of New England (and neighbouring areas) happen to carry, in itself does the mice no harm at all. The deer ticks really flourish when there are lots of mice around (which happens when the acorns get the right weather conditions for making lots of acorns). The female ticks gleefully feed on the mice, and when they are engorged with blood they become huge—as you can see from the photo on the next page. The tiny little puny thing is a female tick before lunching on a mouse. The huge fat tick is the postprandial version—a female who has engorged herself with meals of blood and has started producing eggs, which you can see on the lower part of her abdomen. (It seems that deer ticks are not very interested in anything after they have produced their eggs, so a Jenny Craig weight-watchers regime for overweight ticks would have no subscribers.) When she is engorged and gravid with eggs, the female tick crawls up a blade of grass and waits for the next host to come by. It's usually a deer, but unfortunately, sometimes it's a human.

Under Arrest. Dr. Ostfeld collars one of the usual suspects: a mouse detained while transporting ticks across the forest floor.

And that's the problem—with our wonderful adaptive behaviour strategies we have explored all manner of environments that we weren't originally designed for—and often when we crash into a new world, we

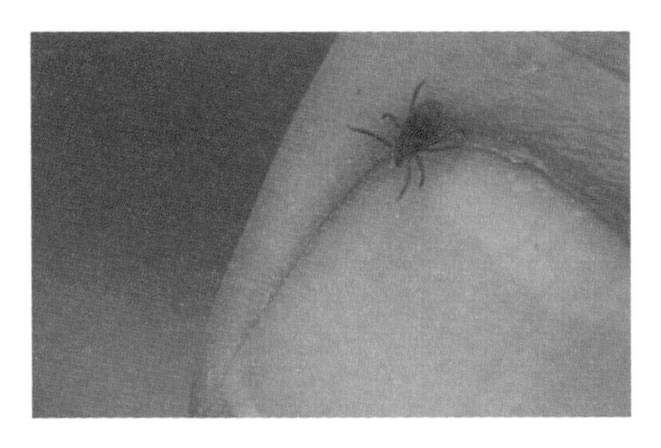

An Idea of Size. Ticks are very small and can often go unnoticed. That is why a thorough tick check of yourself is important after travelling through any backwoods areas.

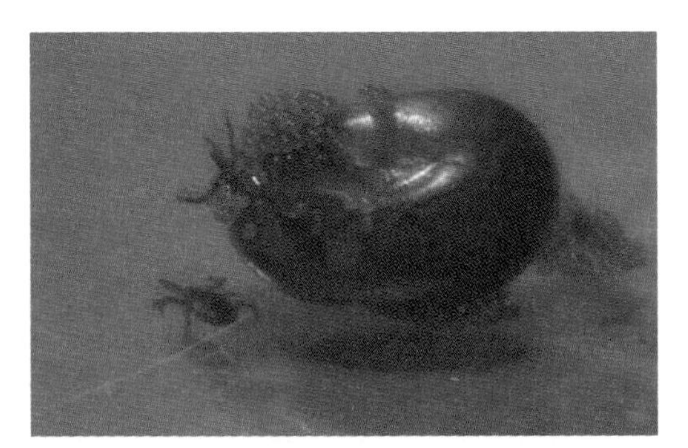

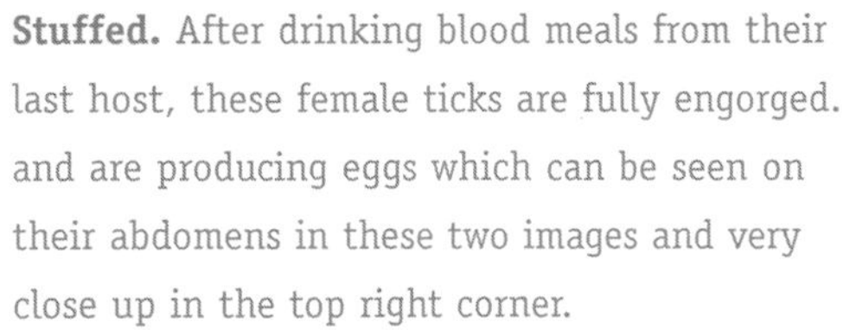

Stuffed. After drinking blood meals from their last host, these female ticks are fully engorged. and are producing eggs which can be seen on their abdomens in these two images and very close up in the top right corner.

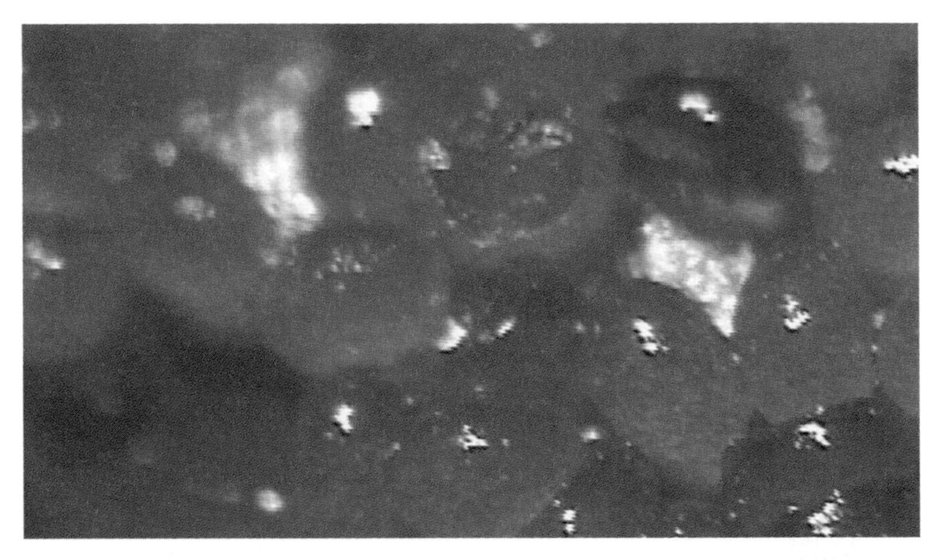

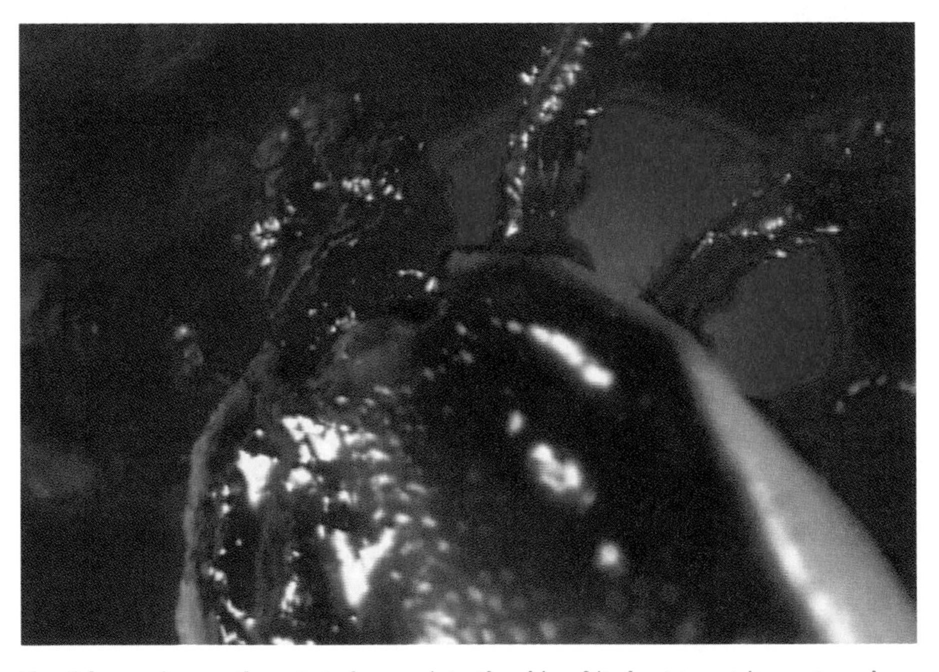

The tick uses its mouth parts to burrow into the skin of its host to get its next meal.

(Opposite)
How Your Dog Ticks. This is the American Dog tick. It looks exceptionally ferocious, but (a) it isn't and (b) the dog never sees it this big.

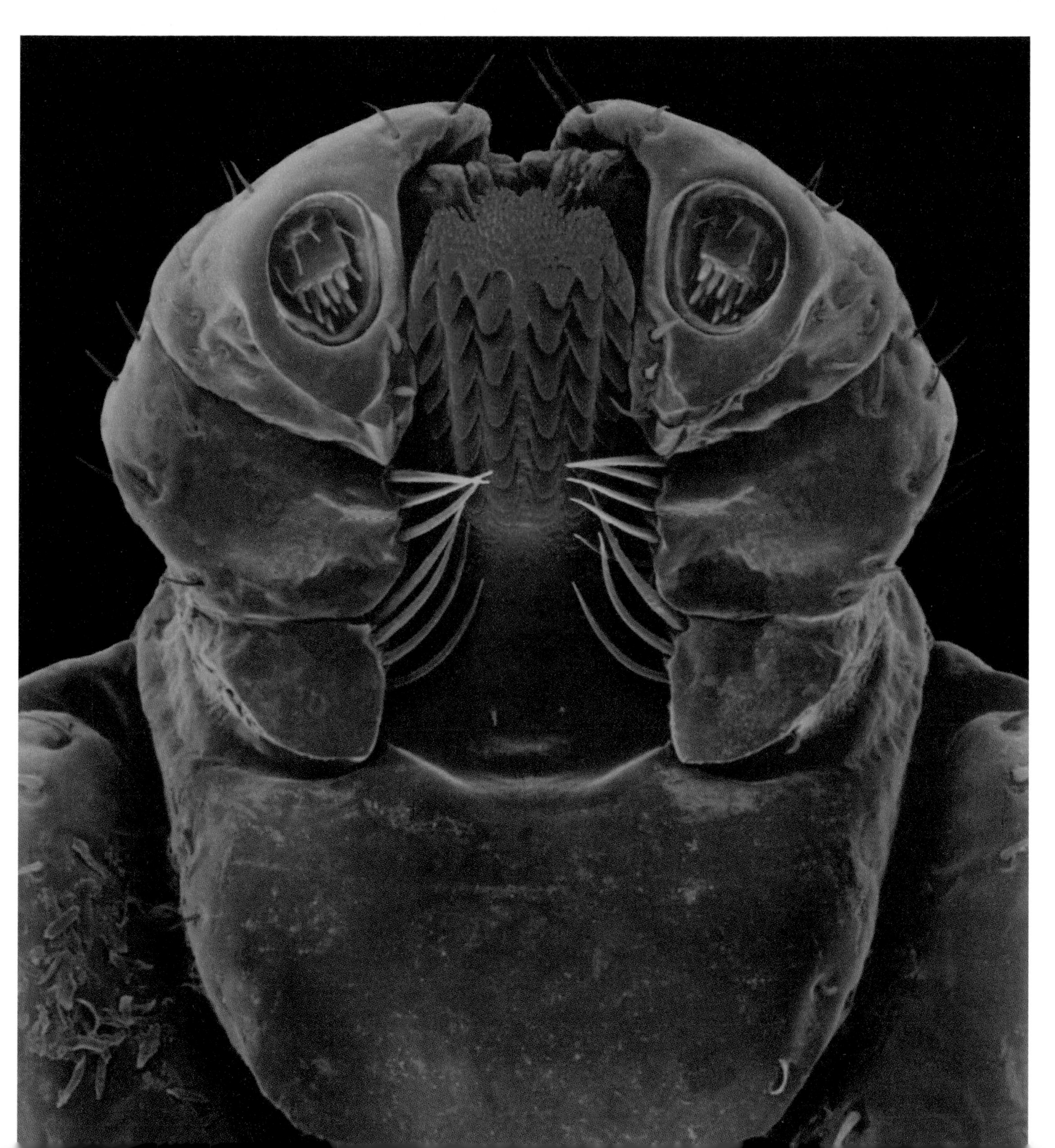

Smart Enough. . . Knowing a little bit about the deer tick and how to do a proper tick check after a walk, means you really *can* enjoy a walk in the woods.

pick up a new disease. In this case, the tick lands on us and over a process of a few hours bites us. The bite contains—among other ingredients—the *Borrelia* bacteria, which gives us the skin rash and arthritis and other problems that make up the illness we call Lyme disease.

This book is not necessarily the best place to discuss Lyme disease, but it is a particularly nasty and insidious illness. As a patient who has suffered severely from Lyme disease, Lia McCabe explained that it really drains a person's *joie de vivre* and leaves them feeling extremely ill with painful joints and low-grade fevers. Nowadays—fortunately—more and more physicians are aware of the diagnosis, and treatment (with the appropriate antibiotic) is likely to be started relatively early. Even a decade or so ago, it was a diagnosis that, because of the disease's rarity, was often missed, and so the illness would drag on and on as it did in Lia's case.

Lia's take on the situation is wise and helpful: her main objective is to be smart enough to enjoy a walk in the woods. And that means knowing what could happen and taking sensible precautions to ensure that it *doesn't* happen.

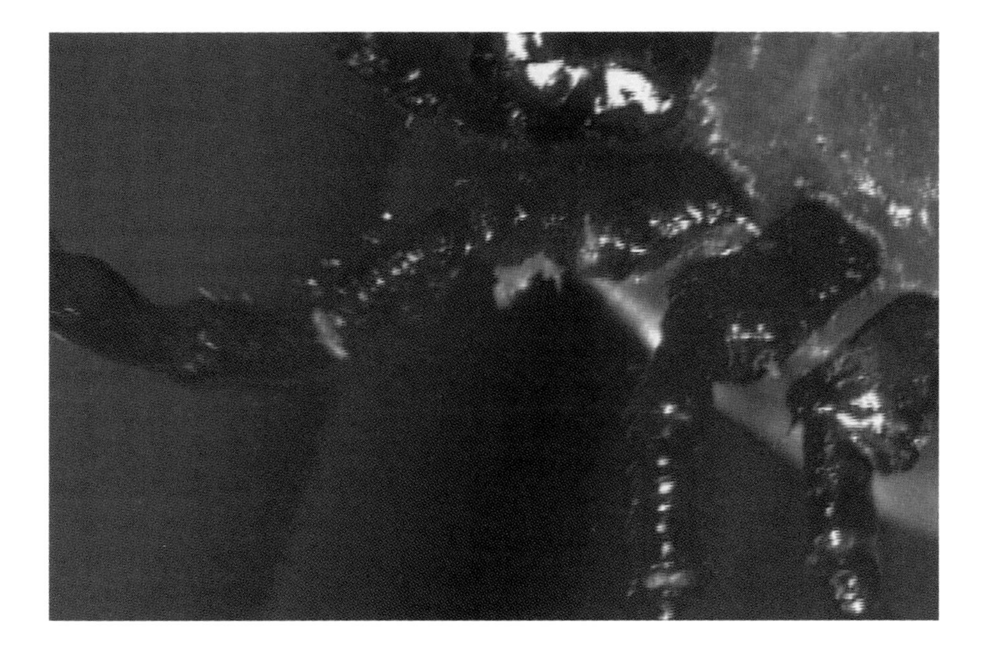

Practical Points About Ticks & Lyme Disease

1. Don't be frightened about going for a walk in the woods—or camping.
2. Use insect repellant.
3. Ticks take several hours to embed into your skin and transmit infections, so when you get back from a walk or when you wake up after sleeping examine your whole body in a well-lit area and look very carefully for anything that looks like one of the creatures you see in these photos. That's what we mean by a proper TICK CHECK.

It Figures—It's Chiggers

If there is such a thing as a free lunch (even though the old adage tells us there is no such thing), for the chigger mite, it's us humans. We are their huge, generous, munificent, bountiful, and totally free lunch. All they have to do is to hop on board and eat until they can eat no more. Then they drop off and have kids who grow up to do the same thing. From every point of view, chigger mites are a real pain.

When they get onto your skin, chiggers, unlike most ectoparasites (i.e. the creatures that cling to the outside of us, as opposed to the things that make their home inside us), don't burrow in—they just push their mouthparts into the skin and start feeding. They have evolved a method of injecting a chemical mixture into the area that allows their pathway to our blood and tissue fluid to remain clear—and, unfortunately from our point of view, this mixture causes us pain. Very great pain. In fact, it has been said that weight-for-weight and size-for-size no living creature can cause as much pain as the chigger.

They start out as little larvae—known as nymphs—and mature into parasitic forms that then leap aboard a passing person, find a nice area of skin, and sink their heads deeply into the human lunch bucket. We get a severe burning pain and they get lunch with full board and lodging for four days or more. When they have had enough, the chiggers quietly drop off our bodies, lay eggs on the ground, and start the whole cycle again.

Up Tails All. A group of chigger mites chow down.

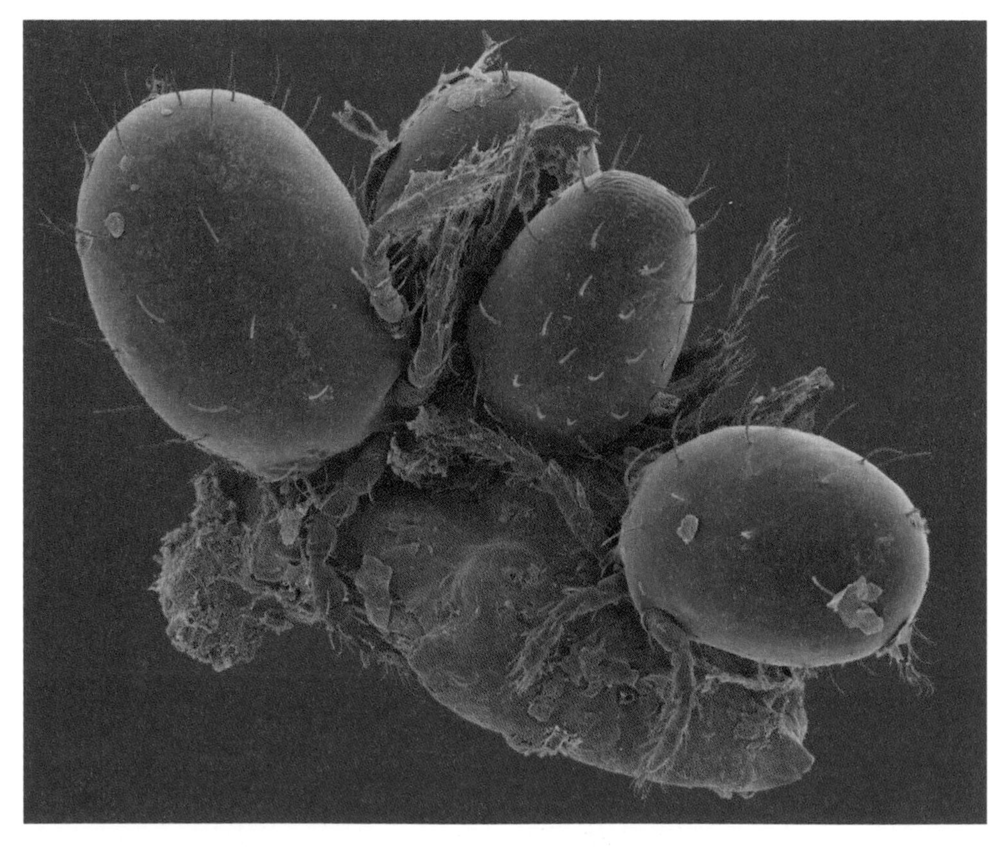

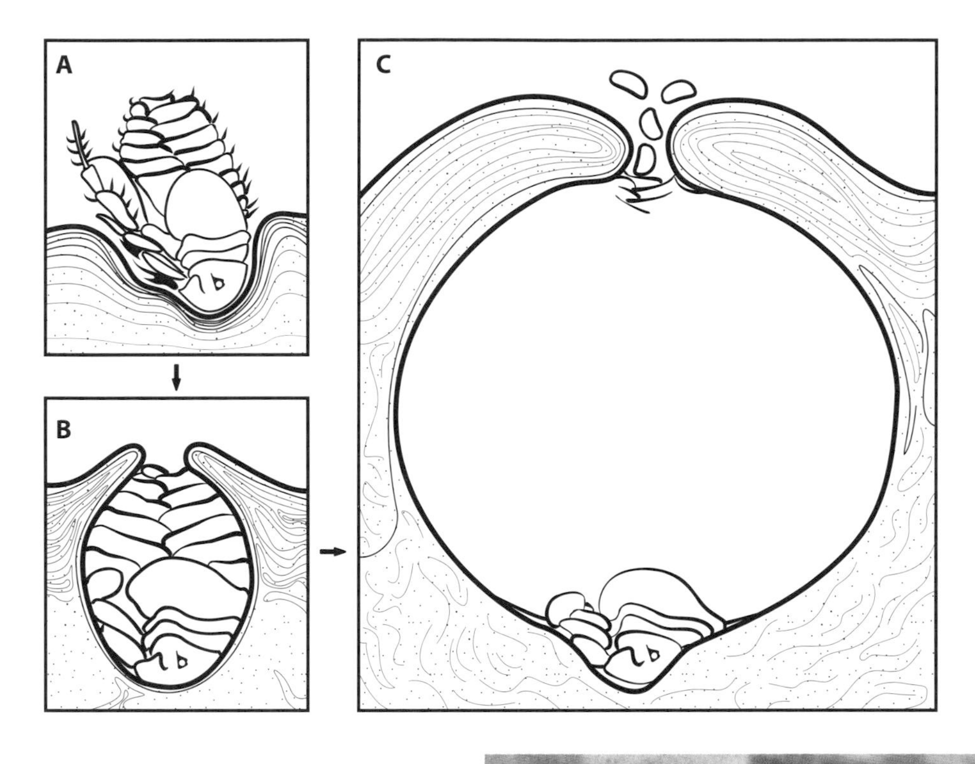

I've Got You Under My Skin. (Artist's interpretation of the chigger mite's theme song.)

"There's Your Problem. . ." Separated from the greatly relieved patient, the chigger mite's cystic area looks as excruciatingly painful as it was.

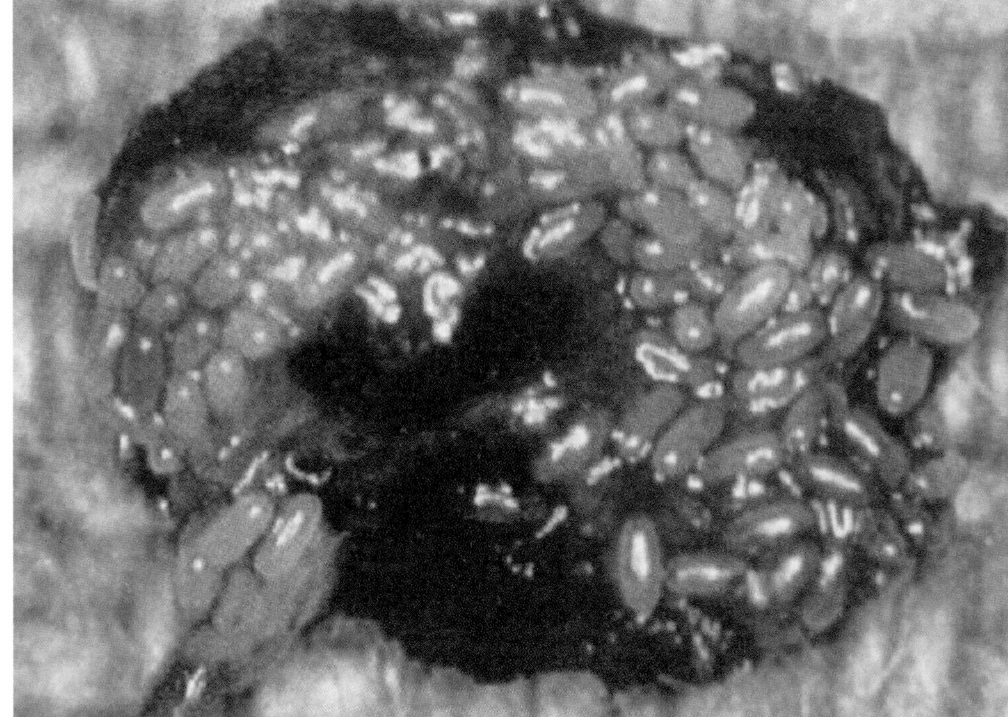

For us, a red, itchy, painful bump on the skin is the only memento they leave behind them.

Fortunately, chigger mites don't spread any known disease, and even more fortunately, they don't do very well at temperatures below about 60 degrees F. Which is perhaps one of the main advantages to living at the North Pole.

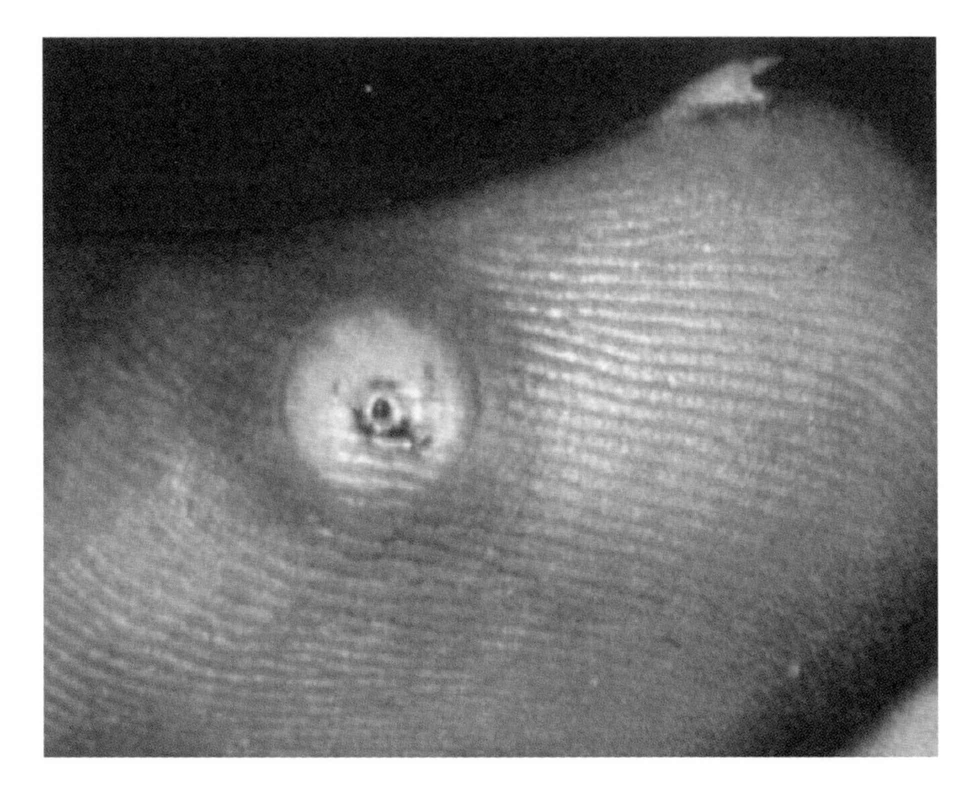

All the Better to Eat You With. Using its mouth parts a chigger mite will drill a hole in your skin and inject a slew of chemicals that help it get its food out of you and help you develop a painful, itching weal, as seen in this picture.

Man's Best Friend. One of the mammals in this picture has been called a real fleabag. The other is a dog.

Flea for Your Life

This is Buddy, a Pomeranian dog who has clearly just been to the canine hairdresser. As you can see from the relative positions of the dog (who is clearly having a very good hair day) and the human (who is clearly not), dogs of this size and shape make very good lapdogs. What you probably did not know is that the whole idea behind the lapdog concept may have much to do with fleas.

Fleas are pretty good at improvisation. Although they tend to specialize—dog fleas prefer dogs, rabbit fleas prefer rabbits, and so on—when there aren't enough specimens of the correct host around, fleas will settle on the next best thing. For many centuries we humans have realized that quite a high proportion of human fleas will wander off onto any warm mammalian body that is close at hand. Or at lap, for that matter. Hence—it is thought—the practice of sitting with a miniature-sized dog on your

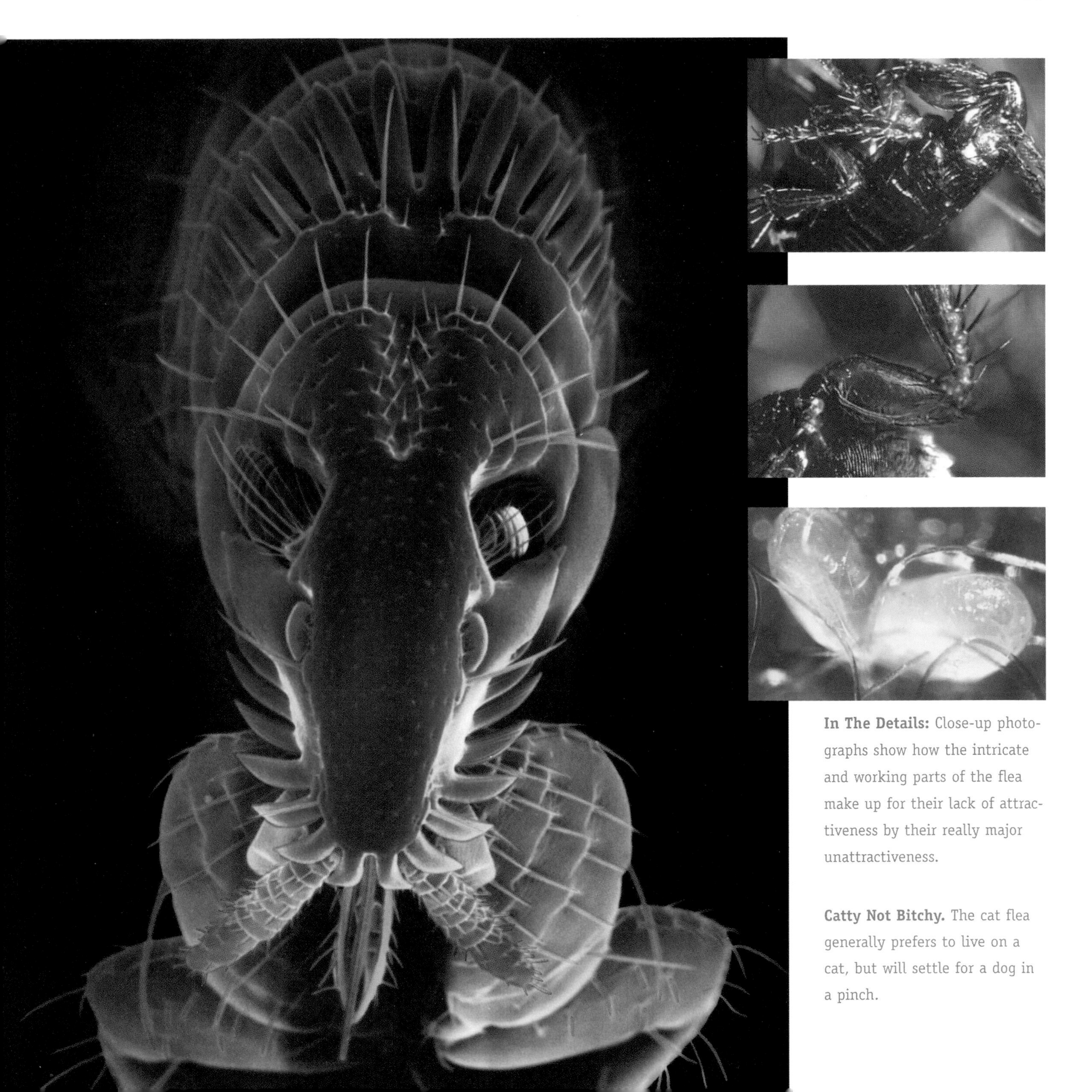

In The Details: Close-up photographs show how the intricate and working parts of the flea make up for their lack of attractiveness by their really major unattractiveness.

Catty Not Bitchy. The cat flea generally prefers to live on a cat, but will settle for a dog in a pinch.

Dr. Terry Galloway

Trick or Treat. This face looks like a Halloween mask, but it is actually a flea seen from much closer up than is normally necessary.

lap may have served the helpful function of allowing some of the human's flea population to migrate onto the dog. It is believed (and I'm not kidding) that this phenomenon is what gave rise to the slang term "fleabag."

On a much more serious note, this trans-species migration of fleas can have major and serious consequences for our species. There is some evidence to suggest that the outbreak of the Black Death—bubonic plague—began when the population of rats suddenly decreased due to disease, and the fleas jumped ship from rats to humans, bringing their unwelcome cargo of plague bacteria with them.

High-Speed Hitch-Hikers

Fleas are extraordinarily efficient hitchhikers. In a famous experiment, scientists released three rabbits into a two-acre field that contained 270 specially marked rabbit-fleas. (I'm afraid I have forgotten how the fleas were marked. Perhaps the scientists clipped tiny metal tags to the fleas' ears, although this would have been technically difficult since fleas don't have ears—as you can see from this photo.) The outcome was utterly astounding: virtually every flea ended up on a rabbit. Seen from the fleas' point of view, this would represent—according to some rather rough calculations—a feat of derring-do equivalent to jumping onto a truck travelling at 400 miles per hour through a forest about 125,000 miles across, almost half-way to the moon.

The fleas' ability to home in on the passing mammalian traffic depended on amazing powers of detecting warmth, vibrations, and carbon dioxide. When those three signals occur together, fleas have learned that (or rather, "have evolved to respond appropriately because it means that") a suitable host is in the neighbourhood and it's time to move. So if you're in a flea-infested area and don't want to get bitten, try not to give off carbon dioxide, stop vibrating, and—if at all possible—go into a hypothermic coma.

For Dr. Terry Galloway, a talented and enthusiastic entomologist in Winnipeg, the world would be a poorer place without fleas because even though they are not particularly loveable or huggable, they are, in their own way, sort of cute. Were they scaled up to human size they would be superheroes. They would be capable of leaping a height of perhaps 120 feet and—this is significant—they would be able to make love for about eight hours. As Terry described it, the male flea and the female flea literally lock into each other. I wasn't quite clear about the anatomy, but, basically, one of the fleas is somewhat channelled like a tram track on the abdominal side and the other one just sort of... well you get the idea. And there they stay for hour upon hour, just being together. No dinner beforehand and no cigarette afterward.

When you look at the female flea close up and try to imagine being locked in an embrace with her for eight hours or so, perhaps it underlines the truth of the old saying "laugh and the world laughs with you—cry and you cry alone."

Trying Not to Let the Bedbugs Bite

Bedbugs are the cat burglars of the parasite world. They are so efficient and so sneaky at their job—which consists of biting you unnoticeably and drinking a very small amount of your blood—that you can be bitten all over your body during the night and not wake up for an instant. They're single-minded (well a bit less than that really, since they don't have a mind as such) and their bite is subtle and painless. They inject anticoagulants so that the tap (your blood vessel) doesn't clog, and they also inject anti-inflammatory agents so that you don't get a pimple or a major irritation that might wake you up and cause you to scratch yourself, thus ending their feast.

Not only do bedbugs not cause you any disturbance, they don't carry any known diseases. In many respects, the bedbug is the ideal ectoparasite. Ugly as all get-out, but a leader and a role model among ectoparasites.

The view from above and below. . .

Louse-Y Luck

As almost everybody knows, there are three distinct species of lice—the head louse, the body louse, and the pubic louse. I must admit I have always dimly wondered how that came about. I had a vague idea that body lice evolved somehow and then wandered about all over the body before finally making their home in the chest or armpit. In fact, it was the pattern of human evolution that probably started the whole process. Early in our history—most likely before the era of the Neanderthals—our bodies were almost totally covered with hair (apart from the eyeballs and a few other orifices, I imagine). At that time it seems likely that there was only one species of human louse and it probably wandered wherever it wanted. Over the millennia, human beings evolved a fascinating

(Opposite)
Auditioning for a Horror Movie. Looking like an extra for the next sci-fi Hollywood monster extravaganza, this bedbug *Cimex lectularius* is actually very good at biting you without you noticing.

The louse at home, in your hair.

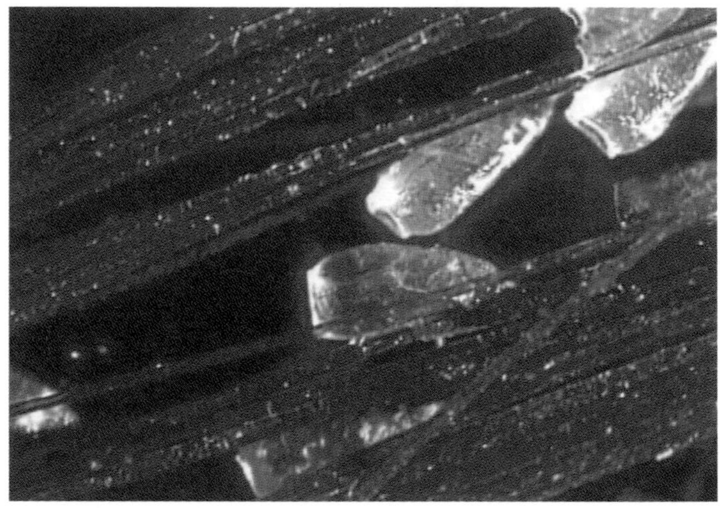

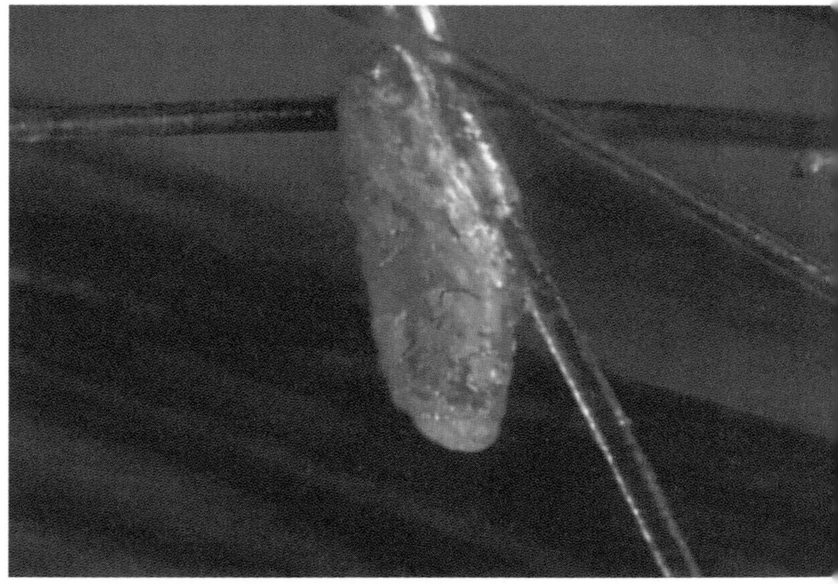

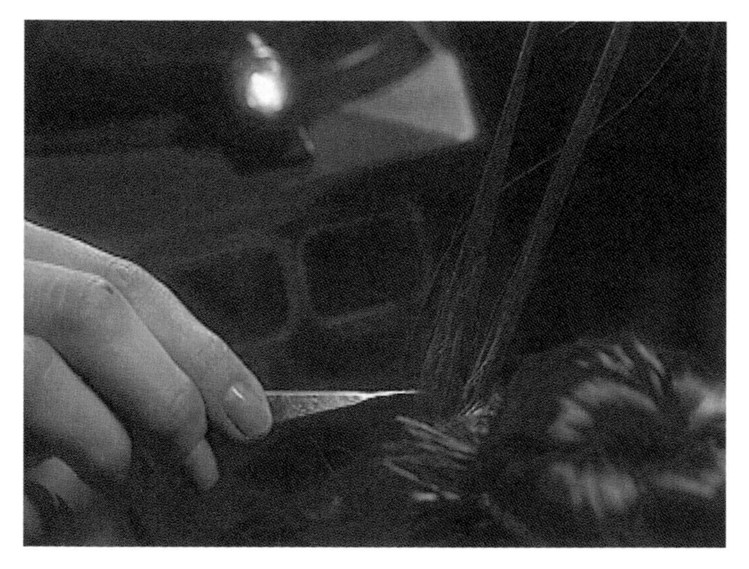

A Nitpicker's Work Is Never Done. It's a painstaking process in which every hair is checked for nits and one by one they are removed.

(Right)

The Root of the Problem. Nits, nits, and more nits. Each one is waiting to hatch into a new louse and another generation of pests.

and idiosyncratic pattern of skin decoration—regional baldness. We humans became hairless in various places (the face and most of the back, for example). We became naked apes in patches, leaving little islands of hair on our heads, on our chests and armpits, and in our pubic areas. Faced with barren deserts between these hirsute oases, the lice evolved right along with these changes and became adapted to the various types of hairs in the three districts. Therein lies the key to three different species.

The head louse is cunningly adapted to hold onto thin and closely spaced hairs. It is so good at clinging on despite the host's attempts to scratch it off that it spends most of its time there and glues its eggs—called nits—to the shaft of the hair with a cement of its own making (and presumably the subject of several patents). The eggs hatch, and the nits crawl out and start the whole process again—unless you do the routine with the nit comb and the special shampoo or employ the services of a specialized person to do the removal for you—the professional nitpicker. (Hence the use of the word as a metaphor for a person who delights in minute and detailed and repetitive work.)

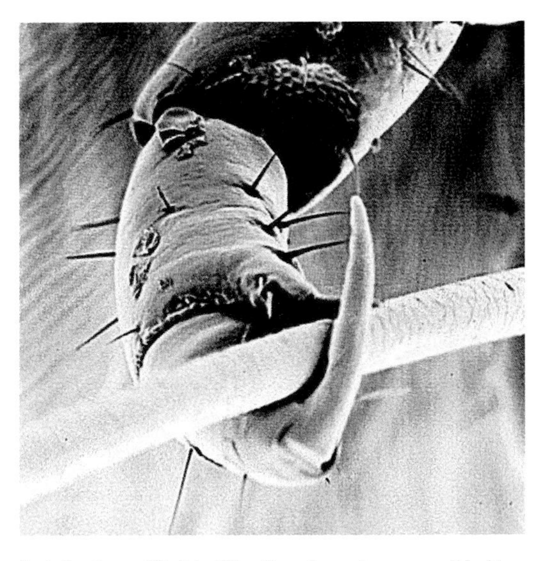

Latched on Tight. Whether hanging on with its pincer (above) or cemented onto a hair, like the nit shown below, the louse is well adapted to preventing its hosts from removing it with any ease.

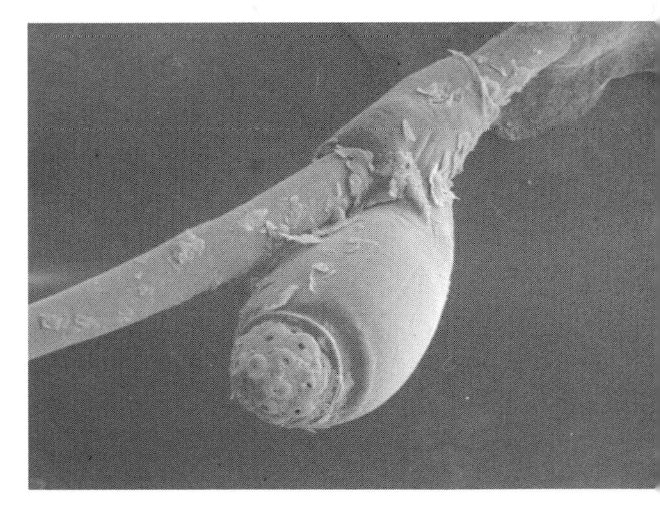

The body louse, on the other hand, is adapted to the wider spacing of the slightly coarser body hairs and can spend a fair amount of its time (and yours) inside the seams of your clothing rather than attached to your hairs themselves. That is why body lice are a real problem in times of emergency (such as war). When we humans are prevented from changing and washing our clothes regularly, the lice can establish a secure home. In the past, the presence or absence of lice in armies actually made the difference between victory and defeat. There is some evidence to show that Napoleon's armies, for example, owed part of their success to a rigid and very strict policy of delousing and uniform-changing, which ensured that the soldiers were free of lice and free of the several diseases (including typhus) that the lice carried.

Finally, there are pubic lice, adapted to clinging on with unbelievable tenacity to the thick and curly pubic hairs. As you can see from the photo on the next page, they have exceptionally well-developed and tough claws. There seems to be a divergence of views regarding the exclusivity of their haunts. Some experts believe that pubic lice can live not only in the pubic hair, but also in other areas of thick hairs, such as eyelashes. Others believe that pubic lice only get to those areas in the context of intimate contact.

Full Frontal Rudity. This is the pubic louse—and as you can see its tenacity is related to the power of its claws in clinging on to one's short and curlies.

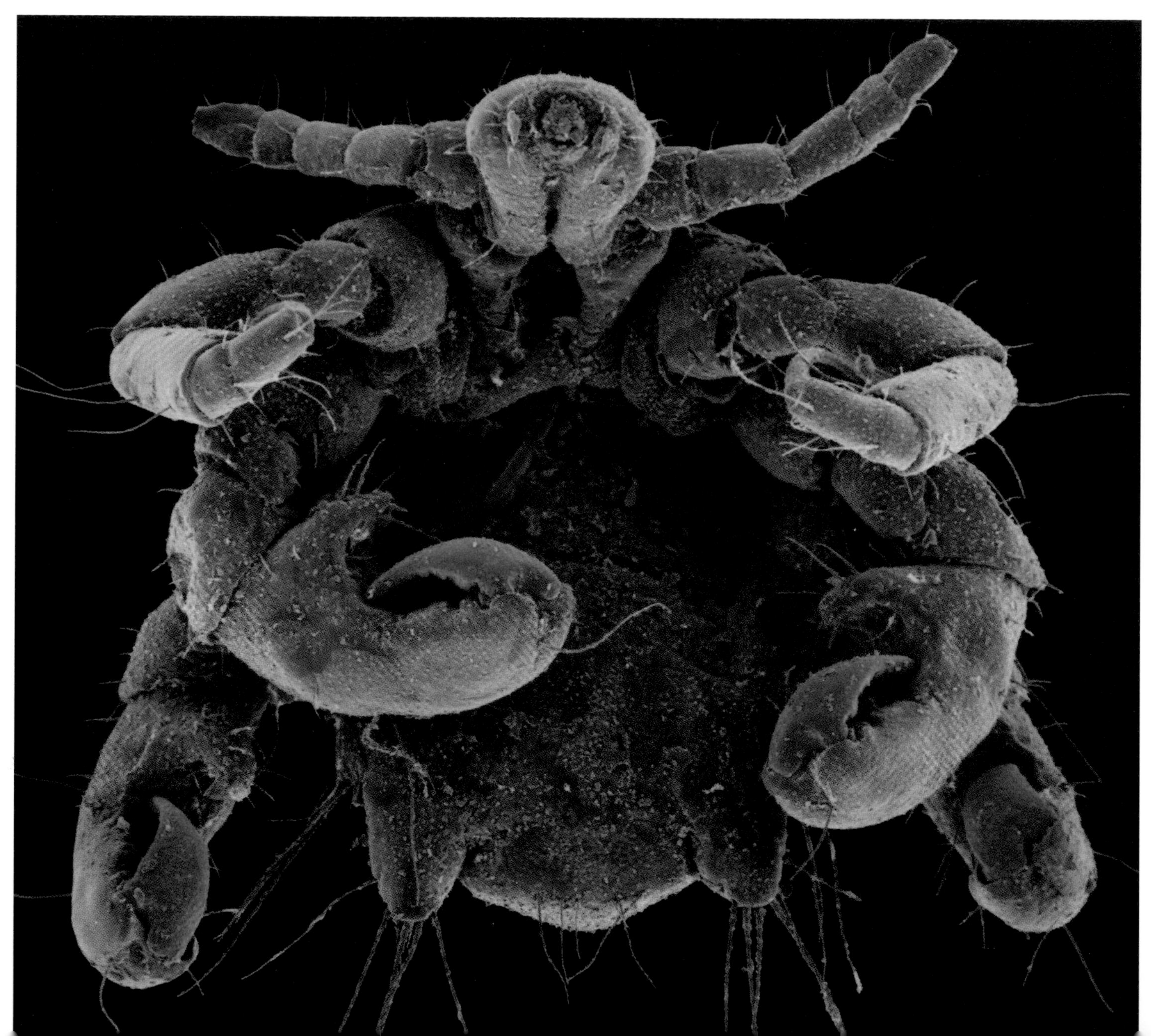

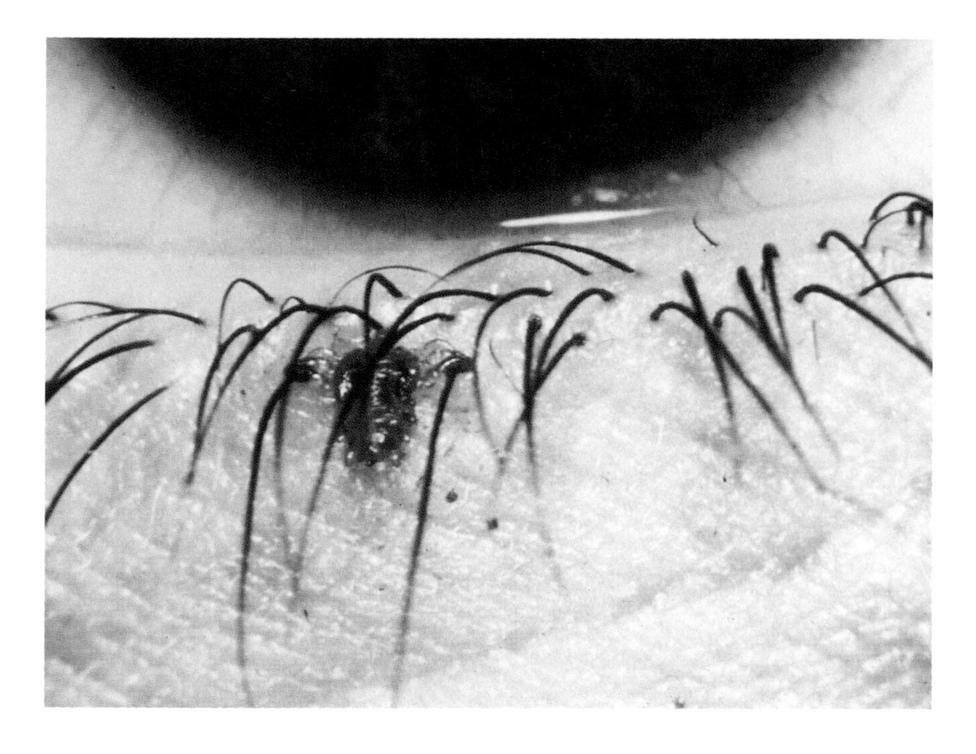

Eye Opener. The louse in this photograph doesn't really belong here. These are eyelashes and the louse is a pubic louse.

Things that Give Pause for Contemplation

If you have pubic lice in your eyelashes, you probably earned them.

Roger Knutsen, *Furtive Fauna*

In Your Face

For thousands of years, we have been fascinated by the human face—sometimes somebody else's, quite often our own.

Bringing Matters to a Head

The top end of the human body is relatively large—compared to the head size of other mammals—and that is primarily because the skull needs to be quite big to contain a brain that has evolved into an organ of about 1,400cc. This development apparently occurred early in our modern history—about 350,000 years ago (and probably on a Friday if most people's week is anything to go by). The shape of the head is directly related to the capacity of the skull to hold the (relatively) large brain within it. Just spend an afternoon watching an ostrich if you doubt me.

Anyway, the head is—by convenience—the residence of the brain, and therefore needed to be quite big for that purpose. It is also the location of some of the most important sense organs. It was advantageous, one assumes, to have the brain in the head and to have the eyes, nose, mouth—with its sense of taste—and ears all within handy reach. We are now totally accustomed to that idea, but it need not have been so. In theory, it would be possible to have the brain in one place and all of the major sense organs somewhere else, connected by nerve fibres and running, as it were, by remote control (or telemetry, to be more accurate). There are few examples of this in biology (although many in aviation, space aeronautics, and TVs). In fact, the only example in biology that comes to mind is that of some dinosaurs—the brontosaurus among them, I believe—which had a fairly big collection of nerve cells in the spinal cord near the base of the tail. This collection wasn't really big enough to merit the title *caudal brain* (a brain at the tail end of the animal), but it certainly was a caudal *ganglion* (a largish collection of nerve cells), and it allowed information from the rear

Reasoning *a posteriori.* The brain doesn't have to be at the head end, although as several dinosaurs found out, things work better that way.

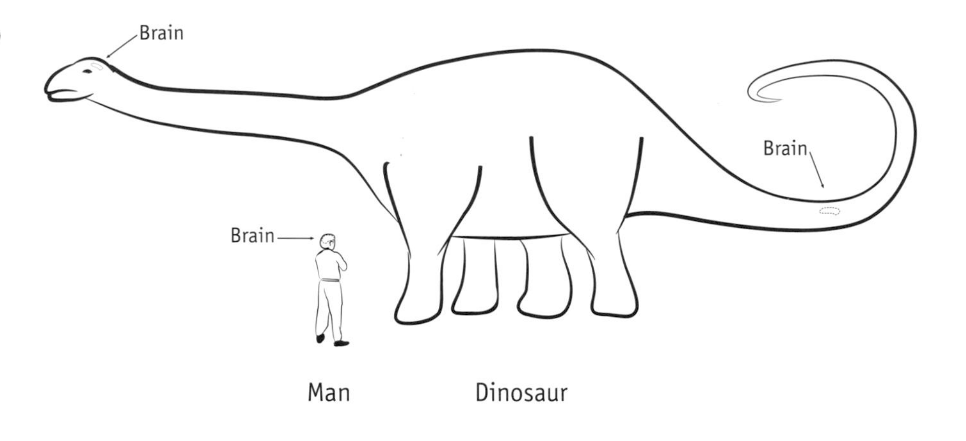

legs to be processed on site, saving the several seconds that would otherwise have been wasted in transmitting the nerve impulses to the head.

The discovery of the caudal ganglia in dinosaurs caught the public's imagination, and many speculated on the two brains as a method of being able to reason *a priori* and *a posteriori*. In my opinion, all it means is that in an emergency, a dinosaur would be unable to distinguish between migraine and hemorrhoids (and we all know someone like that, don't we?).

At any rate, mammals evolved with only one brain, and steadily developed bigger and bigger heads and advantageous locations for their sense organs. That in turn dictated the shape of the human end—and the front edge of it, which we call the face. As it happens, the face is one of the most interesting and complex geographical areas on Planet Human, not only from a neurological point of view, but also from the point of view of its wildlife. The specialized sense organs and the necessity for positioning some of the major portals there (for example, the mouth and nose) has led to fascinating adaptive lifestyles of the bacteria and the other organisms that call our faces home. So let's start by putting our biological money where our mouth is.

Soon to Be a Major Biofilm

I must be honest—before I started the research for this book, I had never heard the word "biofilm." I had no idea that different species of bacteria and other types of wildlife could live together in a genuine community, providing things for each other that they couldn't get for themselves. I also had no idea that the plaque on my teeth was biologically the same sort of thing as pond scum. Although there has always been the occasional Sunday morning when it felt like that.

To put it simply, the human mouth is actually a delicate and highly complex ecosystem in its own right, and a really good place to start looking at how it all works is the most visible feature in the mouth, the teeth.

Teeth are highly unusual structures and represent a genuine triumph in evolution unsurpassed in structure and function by anything, including dentures. From the point of view of the ecology of the mouth, the really significant feature is the tooth covering, a material that we call enamel. The cells of the growing teeth lay down the enamel, covering the developing tooth bud in it, and then remain inside it, protected under the impermeable flawless shell that is totally resistant to most invaders (other than things like fists and hockey pucks, of course). From the point of view of the wildlife of Planet Human, enamel is extraordinary because it is utterly smooth and shiny. Nothing can really get a hold on the surface. To ordinary bacteria it really is like a glass mountain.

At the same time, however, in the average human mouth and on the average human teeth, there is clearly a thriving population of wildlife. It all starts with the particular properties of a bacterium called *Streptococcus mutans,* which has a few superpowers not shared by other bacteria—the main one being that it can cling onto flat surfaces. Furthermore, it particularly likes to do that when there is a film of sucrose (ordinary sugar) around. Sadly, that is one thing we humans are generally quite good at providing. For the last few tens of thousands of

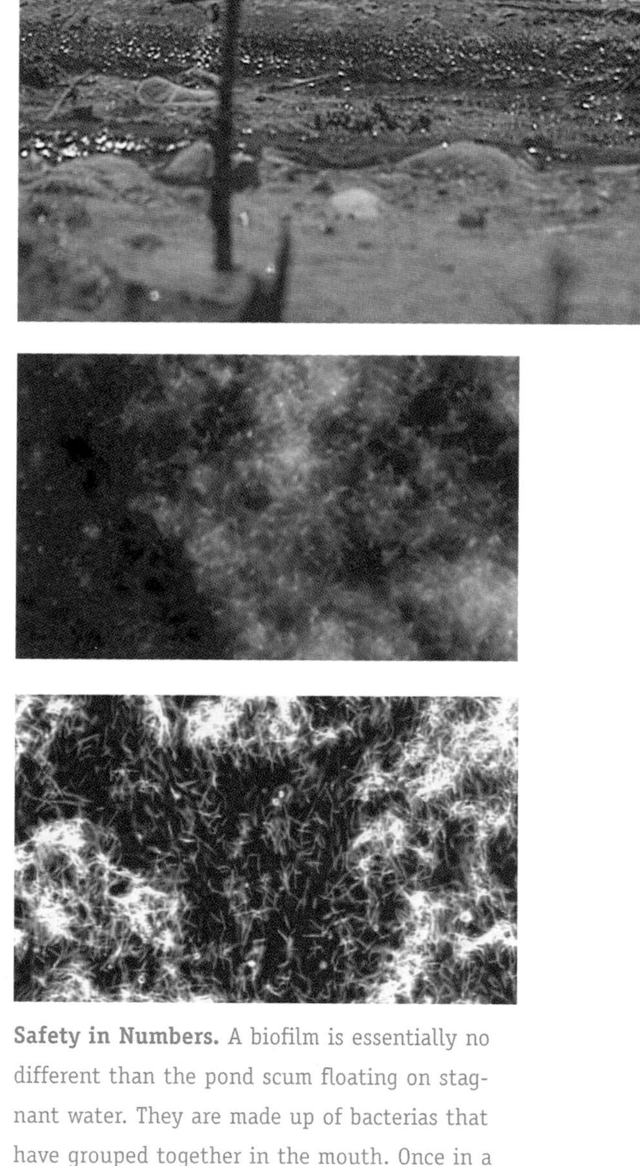

Safety in Numbers. A biofilm is essentially no different than the pond scum floating on stagnant water. They are made up of bacterias that have grouped together in the mouth. Once in a biofilm, together they can usually resist the torrents of saliva that wash most other bacterias and visitors in the mouth down to the stomach and an acid bath.

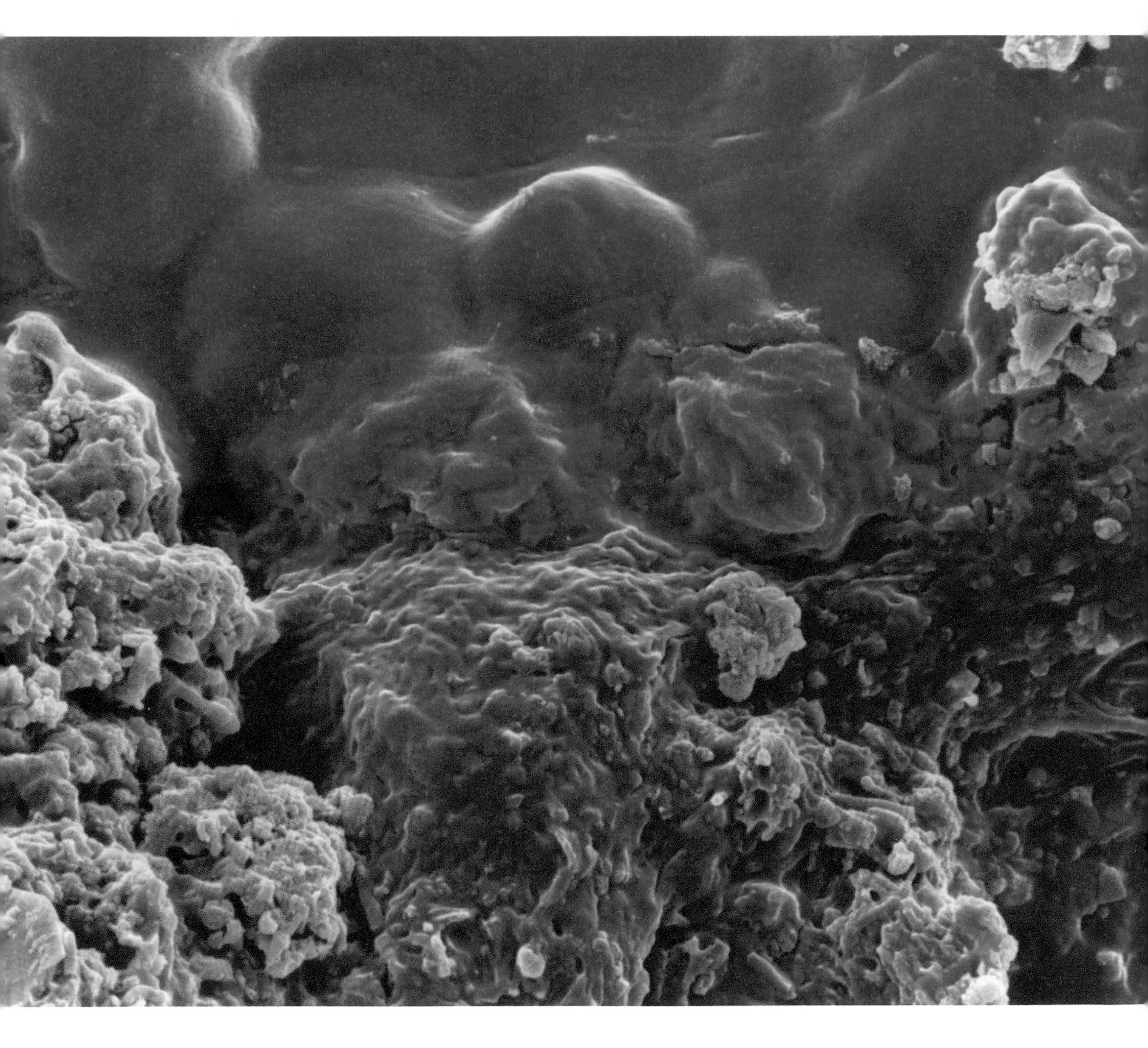

years or so, humans have actively pursued the taste of sugar. Every human civilization has persistently sought the sweet taste, despite the fact that sucrose is, contrary to the widespread belief of all children and most adults, not an actual dietary staple nor a vitamin nor a vegetable. Across the world, human societies have cultivated sugar-producing plants such as beets and canes (and then, of course, we have cultivated large factories to manufacture aspartame and saccharine).

The problem with sugar is not its taste, but the fact that we humans enjoy a slow oral bath of the stuff, and it leaves behind a thin chemical film that the *Strep. mutans* enjoys almost as much as we do. Another of *S. mutans'* properties is that it can produce various enzymes and other chemicals that begin to roughen up the previously glassy surface of the enamel. Which is how the biofilm starts. The carpet of *S. mutans* provides a safe anchorage for other bacteria, and soon dozens of species start to establish themselves in the bacteria equivalent of a growing city. If you're looking at the surface of a tooth, that biofilm will evolve into what we call dental plaque. If you're looking at the banks of a slow-flowing river, we call the biofilm pond scum. The principle is the same (though I am told that very few rivers nowadays can be improved by regular use of dental floss).

(Opposite)
Nature's Clingiest Hanger-on. *Streptococcus mutans* has the ability to fix onto shiny surfaces and, unfortunately for us, other bacteria—like a party of mountaineers following on a rope behind the leader with the crampons—can cling on to the clinger-on. The red areas in this photograph are bacterial plaque on a human tooth, and the rough surfaces are calcified tartar deposits.

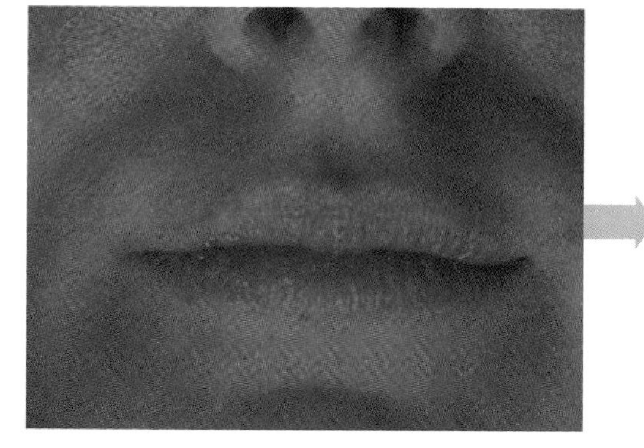

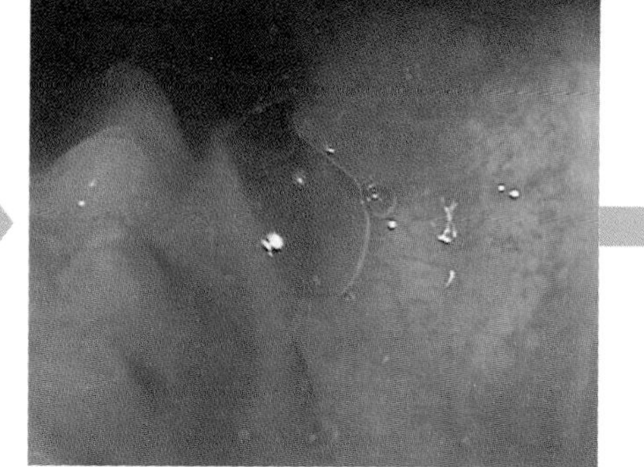

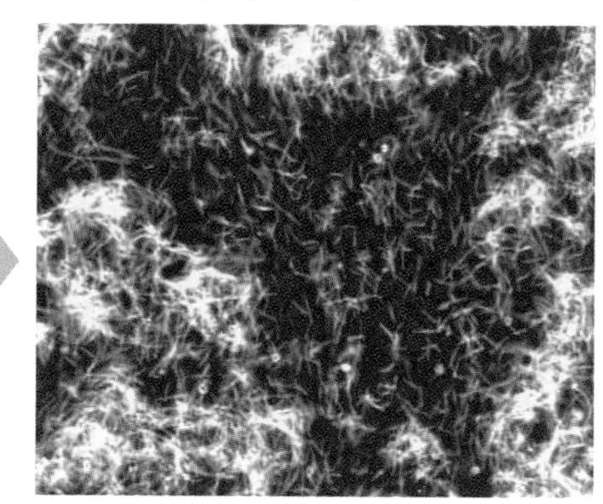

Why Flossing Is Good. Even when the tooth surface looks clear, without proper cleaning the biofilm that we call plaque is right there.

Things I Probably Knew Once But Forgot

The average person produces about 1 to 1 $^1/_2$ litres of saliva in a day (but swallows most of it so it doesn't show on the chin or shirt).

Salivary Salvation

Of course the normal human mouth is not the same as a slow-flowing river. In fact, it's probably more like a rushing torrent for some of the time. From the bugs' point of view, the place is being intermittently sluiced down by fast-moving jets of a solution they regard as dangerous, and even lethal—saliva. As Dr. Julian Nedzelski, Otolaryngologist-in-Chief at Sunnybrook and Women's College Health Sciences Centre in Toronto, showed me, the inside of the mouth is constantly being rinsed with jets of saliva squirted out from three pairs of salivary glands neatly arranged around it. To any unattached itinerant bacteria, this would be like trying to set up a tent at the bottom of the Niagara Falls. So although we create the problem by consuming sugars in the first place, the clean-up crew in our mouths does its best to hose our teeth off with saliva.

Problems with Brimstone

Sulphur-producing wildlife, on the other hand, ironically—although symmetrically—create problems for us at both our top end and our bottom end. If you have recurrent halitosis—a term for bad breath that was actually coined by a toothpaste manufacturer to make it sound as if it were a disease—then it is probably due to bacteria. Halitosis at the bottom end—the characteristic rotten-egg smell of human flatus—is also caused by bacterial wildlife and is an issue we shall grapple with in Chapter 6.

All of us suspect that we have bad breath from time to time. The suspicious signs that we have this problem vary from our friends withdrawing to a distance of three feet away when we speak or exhale, to sudden nausea or vomiting in people who stand close to us, to producing coma or deep unconsciousness in someone we kiss. (There are other explanations for this, I'm told, but I can't think of any myself.)

On a more serious note, some people do have genuine and recurrent bad breath, no matter how many different toothpastes they try and how much mouthwash they bravely strive to gargle-hold-and-spit with. There is a tiny number of people who have rare metabolic problems (there is an actual condition known as "fish breath" in which a very rare chemical is unfortunately produced by the body's metabolism). But for most bad-breath sufferers, the problem seems to recur no matter how much time, effort, and money they spend on dental and oral hygiene Perhaps, like me, you thought the problem was somehow related to dental plaque accumulating in that often-overlooked area inside the lower front teeth or in between the upper molars. Either way, we are wrong. The problem has nothing to do with the teeth, the gums, the stomach, the gullet, or the tongue—either on its undersurface or its pointy, wiggly front.

Lying Tongues. Most of the time what we see on our tongue in the mirror or under the microscope—however alarming the appearance—is actually benign and completely normal.

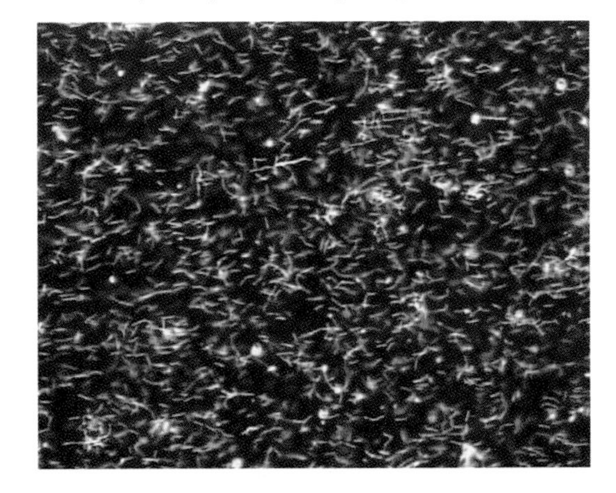

The problem that causes recurrent bad breath is some wildlife that lives at the back of the tongue. In that area, several types of bacteria—species that have not been absolutely identified yet—establish their home and do something that most bacteria cannot do; they break down sulphur-containing amino acids, such as those found in proteins from meat and other animal sources. That activity unfortunately produces a group—well, a cloud, really—of volatile sulphur-compounds colloquially called VSC. These smell, literally, like hell. In medieval times, it was imagined that hell was filled with fire and burning brimstone, the name given to a lava rock

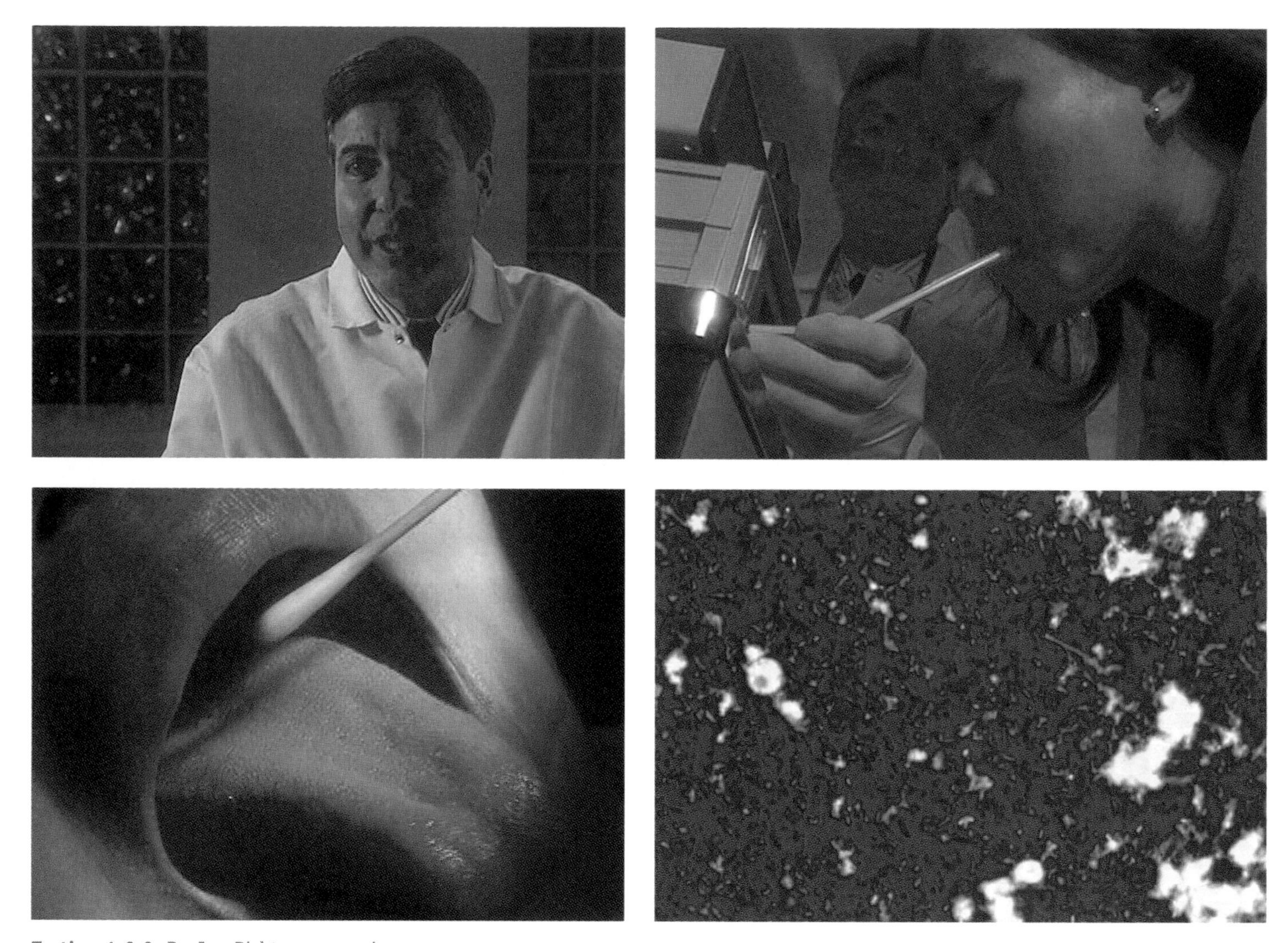

Testing 1-2-3. Dr. Jon Richter uses various methods to test the breath and detect bacteria that foul it up.

rich in sulphur. The smell of hell—as they conceived of it—would essentially have been sulphur dioxide with some hydrogen sulphide thrown in as well. Undoubtedly, it would have been full of VSCs, and dental hygienists relegated to hell would have felt as if they were back at work.

Much of the latest evidence on this subject has been produced and collated by Dr. Jon Richter, an oral scientist and founding director of the Richter Center for the Treatment of Breath Disorders in Philadelphia. Dr. Richter has several methods of detecting the presence of these VSCs—

Things I Never Knew Before

Recurrent bad breath isn't caused by gunk on your teeth—it's caused by bacteria at the back of your tongue producing hydrogen sulphide.

and, by inference, the bacteria that produce them. First of all, when a patient sees him for help and advice on bad breath, he asks them to breathe out and as they do so, he smells their breath. This is important because, as we shall see, quite a few people who labour under the impression that they have perpetual bad breath, actually do not.

Dr. Richter then uses a device called the *halitometer* to measure the VSCs, and best of all, he does something that most oral microbiologists don't do. He uses a swab to take a sample of the material at the back of the tongue (you have to be a bit careful to avoid triggering the gag reflex) and puts the swab into a special fluid of his own devising. The fluid contains material that allows the bacteria to feel at home and do what they do best—produce sulphur. In the mixture is some lead acetate, a colourless liquid. If there are bacteria on the swab that produce sulphur compounds, the products will react with the lead acetate and produce lead sulphide, which is black. So if you've got the bad-breath bugs, the bottle goes black. If you haven't, it doesn't. This test is valuable even if your breath doesn't happen to smell bad because if the sulphur-producing bacteria are there, they will produce bad breath at some stage.

Obviously not everyone who fears that they have bad breath actually does—a lot of people who are socially sensitive do not have these bacteria (or genuine bad breath). But for those who do, Dr. Richter—and many oral hygienists—will show you how to scrape the back of your tongue (taking care to avoid gagging) with a special plastic scraper and/or gargle with a mouthwash containing the chemical chlorine dioxide, which is particularly good at knocking off the bad-breath bugs.

All right, *now* you can exhale.

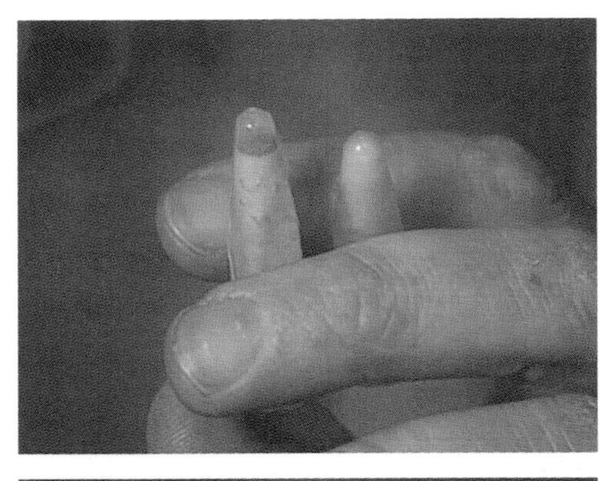

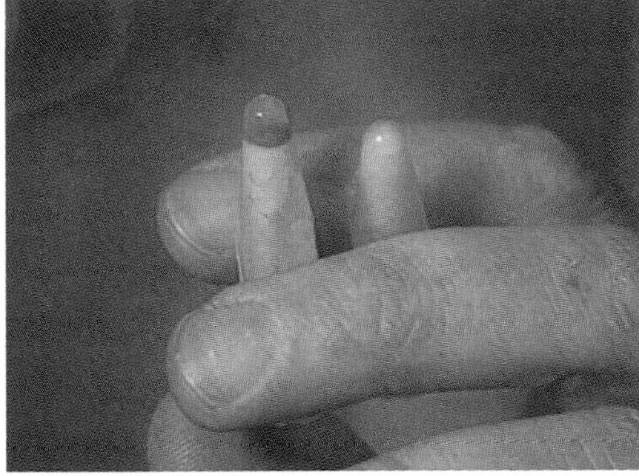

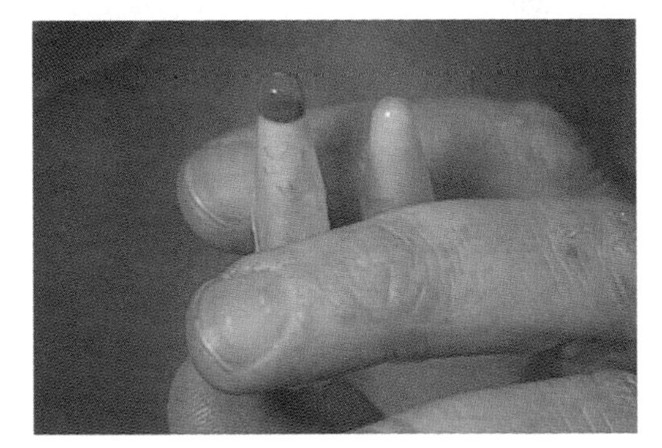

If It Comes Up Black. . . The verdict is clear: if you have sulphide-producing bacteria at the back of your tongue (the ones that are responsible for bad breath), they will turn the clear lead acetate solution in the bottle into lead sulphide, which is black. (And if you don't have them, they won't).

My Enemy's Enemy Is My Friend

So far, this description of biofilm getting a start on breaking through dental enamel sounds like a clear case of bad invaders messing up good human environments, but it is not quite as straightforward as that. Although the biofilm is not at all a good thing for teeth, the oral wildlife that creates it represents, at the very least, a wildlife that is better than some really nasty species that could be living there. If the oral environment can be compared to a city, then the normal oral flora is like the average urban population. Some of them are very well behaved, a lot are a bit noisy, a few drop litter, and a very few scribble on the walls and rip out phones. But in general, the city sort of muddles along.

If, on the other hand, there is a major fire and the normal citizens flee—or are driven out—the place can be occupied by some really law-

Order that Looks Like Chaos. If you were a visitor from another planet, our cities would probably look like totally random chaos—the rules by which they are run are not easy to see. The same is true of the microbiological life forms in biofilms.

Things I Never Really Wanted to Know

The types of bacteria that make your breath smell bad are the same types that make your flatulence smell the way it does.

less crowds, armed revolutionaries, in fact. That is our understanding of the normal oral wildlife—it occupies the ecological niches and, even though it is no friend of dental enamel, it holds the fort and prevents genuinely nasty wildlife from taking over the place. One example of the unwelcome and opportunistic type of occupant is a particular yeast.

Candida on Camera

A very useful way to understand the principle of a balanced and normal wildlife population is to look at what happens when the balance is upset. When we get a bad chest infection, for example, we may be prescribed antibiotics. Quite often, the antibiotics will kill not only the specific bacteria that are causing the chest infection—such a thing is not yet possible—but will also seriously affect normal bacterial populations, such as

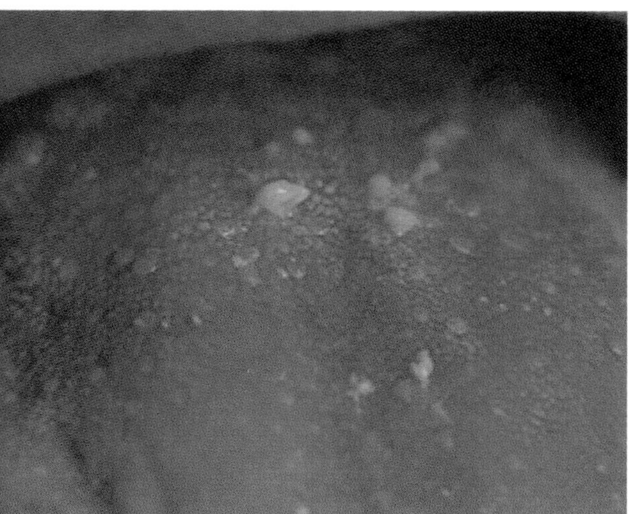

Not as Pure as Driven Snow. These snowball-like things are actually one form of an infection by the common oral yeast or fungus, *Candida albicans*. *Candida* gets a better-than-average chance of establishing itself when the normal bacteria that occupy most of its favourite sites in the mouth have been knocked out by antibiotics.

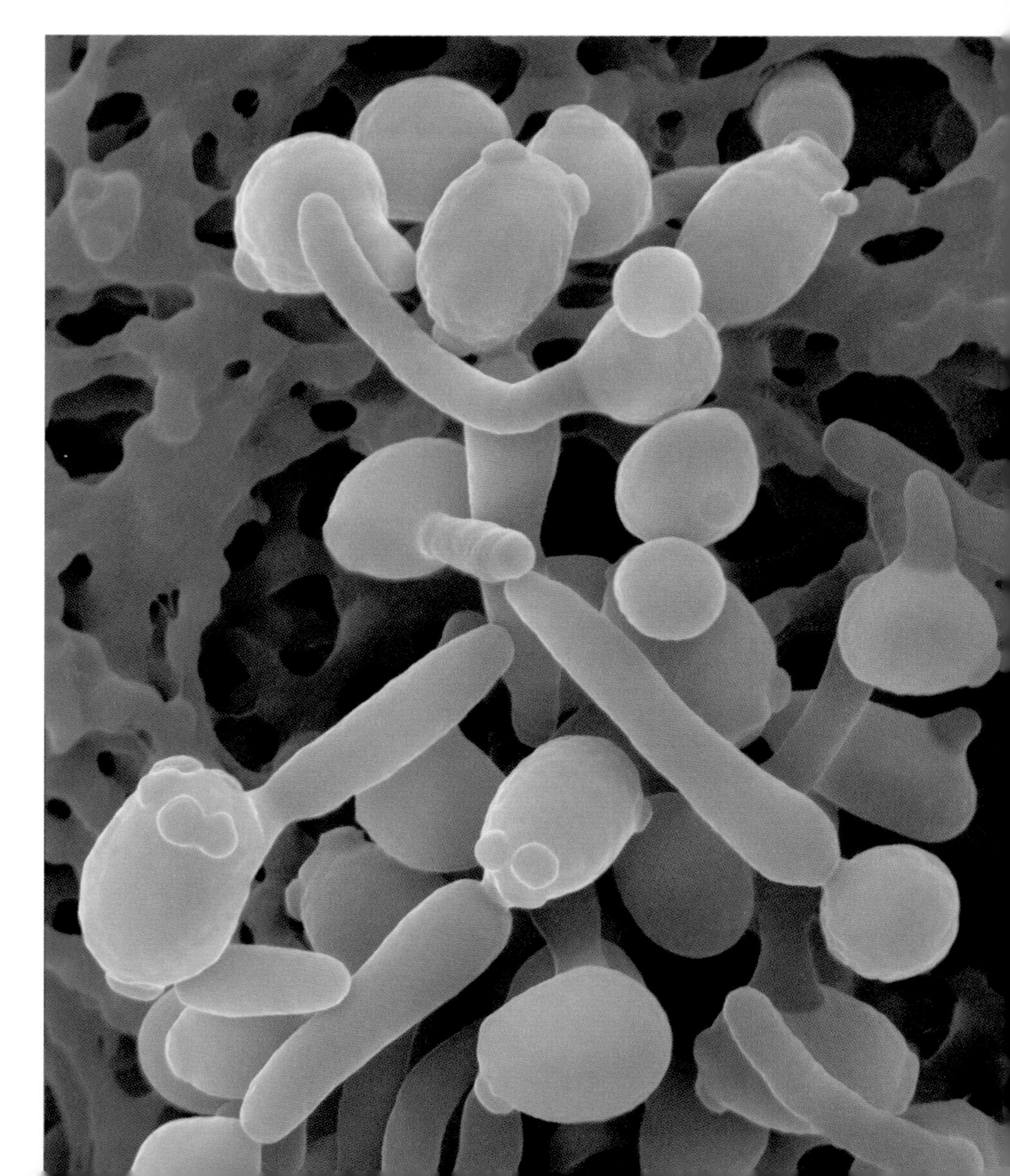

those in the bowel (which is why some antibiotics cause diarrhea) and those in the mouth. A common problem with antibiotics, particularly among the elderly and those with other medical problems, is that the loss of the normal oral flora allows the ecological crevices to become occupied by other organisms, of which the commonest squatter is a yeast called *Candida albicans*.

Candida is always around. It is normally a fairly harmless fungus, a few strands and branches of which sit around in almost everyone's mouth. Since all of the available ecological niches are normally filled by bacteria, the *Candida* cannot do very much except sit there. When antibiotics have cleared out the usual occupants of the mouth, however, *Candida* seizes the opportunity and moves in. The symptoms of a *Candida* infection in the mouth vary in degrees of severity from no symptoms, to a slightly nasty metallic taste, to a sore tongue, to (rarely) difficulty in swallowing, and, in a very few cases, to overwhelming infection of the blood. The treatment is fairly easy nowadays—*Candida* is almost always sensitive to antifungal treatments, but it is an example of the usefulness—although we don't realize it at the time—of the presence of normal oral wildlife.

The Gifts of Motherhood

Although it may seem politically incorrect, there is good evidence that shortly after you are born your mother gives you a starter kit of normal bacteria along with maternal love and mother's milk. All three of these things are absolutely vital—the bacteria no less than the love and the milk.

The man who knows most about this is the energetic and thoughtful Dr. Page Caufield, head of the Division of Diagnostics, Infectious Disease, and Health Promotion at the New York University College of Dentistry. He became interested in the microbiology of the mouth and wondered how a newborn infant's mouth became colonized with normal and good bacteria, and why nasty or bad bacteria so rarely moved in and took over. We certainly don't know all the answers, but there seems to be very clear

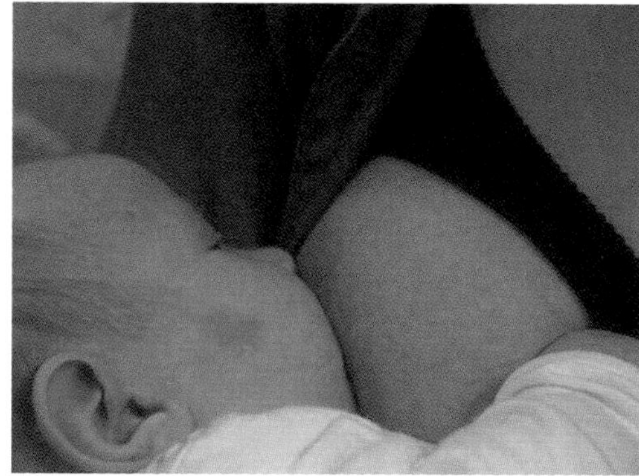

What You Get with Your Mother's Milk. Thousands of paired samples in Dr. Caufield's freezer show that the bacteria that inhabit a baby's mouth are derived from its mother.

evidence that your mother starts you off with the normal species that will occupy your mouth for life, possibly including any propensity to dental decay if you happen to be susceptible. It is a fascinating and extraordinary story—and illustrates several important aspects of the relationship between our species and the wildlife that calls us their home.

In general terms, each of us has a characteristic mix of species of bacteria living in our mouth. The bacteria, for the most part, get along fairly nicely with each other and, for most of the time, don't cause us much trouble either. Dr. Caufield suspected that this stable population of oral bacteria—the normal oral flora—arrived there specifically (and not by mere chance) and that their occupancy served some purpose or function. So he undertook some major research into the question. He took samples of oral bacteria from very young babies, and samples from the mouth and the skin of each infant's mother. Then he compared them. He found that the baby's oral flora did match—almost exactly and almost always—the bacteria from the mother. Although it may be politically incorrect to say so, the father—microbiologically speaking—is almost irrelevant. (So we fathers are going to have to think of things to pass on to our children other than our microbes.) In Caufield's freezer in New York are thousands of pairs of these samples—not just from the US but also from other parts of the world, including Africa and Asia. Wherever you look, the findings are the same: for microbes it's a clear case of "like mother, like child."

This is clearly much more than a coincidence, but what could it mean? Dr. Caufield thinks that this bacterial resonance is actually telling us something very important about the way the human immune system learns what is "normal" in the outside world, and what is "foreign and potentially dangerous." During the first few days of a baby's life, the immune system is "waiting to be programmed"; it is ready to be told what bacteria—and other wildlife—are always around and can safely be ignored. This is the phenomenon of immune tolerance—if a baby's system meets Bacterium X during that phase, it will not develop an immune reaction to Bacterium X the next time it runs into it. Hence the baby's encounter with its mother's bacteria may be an incredibly important part of the education of that child's immune system. As we shall see in Chapter 4 and again in Chapter 6, there are other examples of how the immune system becomes tuned and educated—to the mother's bowel bacteria, and, possibly, also to worms. The conclusion drawn from the work done by Page Caufield and others is telling us something of great importance about how peace is maintained between the different populations on Planet Human.

Why Tears Usually Are Enough

The mouth is, of course, one of the face's major gateways to the interior. But there are others, and although we tend to think of the eyes solely in terms of the visual information they gather, there is a physical aspect that is extremely important in maintaining their function. We need to keep our eyes clear and clean in order to be able to see well. What you see in the photos on the next page, assuming that you can see it clearly and cleanly (meaning that, if you wear contact lenses, you have got them in), is a common sight.

Saving Your Money Where Your Mouth Is. This is a very common manoeuvre—instead of spending hard-earned cash on complicated contact-lens solutions, vast numbers of us use the slobber that nature gave us. Yet, given the complex bacteriological stew that sloshes around inside the mouth, why don't we all get eye infections? The answer is in the magic of tears—they are literally a solution to many problems.

Contact lens wearers frequently use their own saliva instead of the highly complex and rather expensive lens solution sold so readily at the drug store. It is a common (and money-saving) practice, but is it dangerous? Given the fairly large number of bacteria in the human mouth and clinging to the biofilms on the teeth, wouldn't a reasonable person expect that wetting your contact lens with saliva is equivalent to using sewage as a marinade?

You might have thought so, but fortunately nature (in this case anyway) is looking out for us. It so happens that the tears manufactured in our lachrymal glands contain a substance that kills bacteria, as a rather neat demonstration by Dr. Andy Simor, head of Microbiology at the Sunnybrook and Women's Health Sciences Centre in Toronto, illustrates. In an experiment performed for the benefit of this book, Dr. Simor and his team took a fairly concentrated solution of bacteria—actually the *Staphylococcus aureus* bacteria that one would find on a person's skin—and spread it evenly over an agar culture plate (i.e. a Petri dish). After

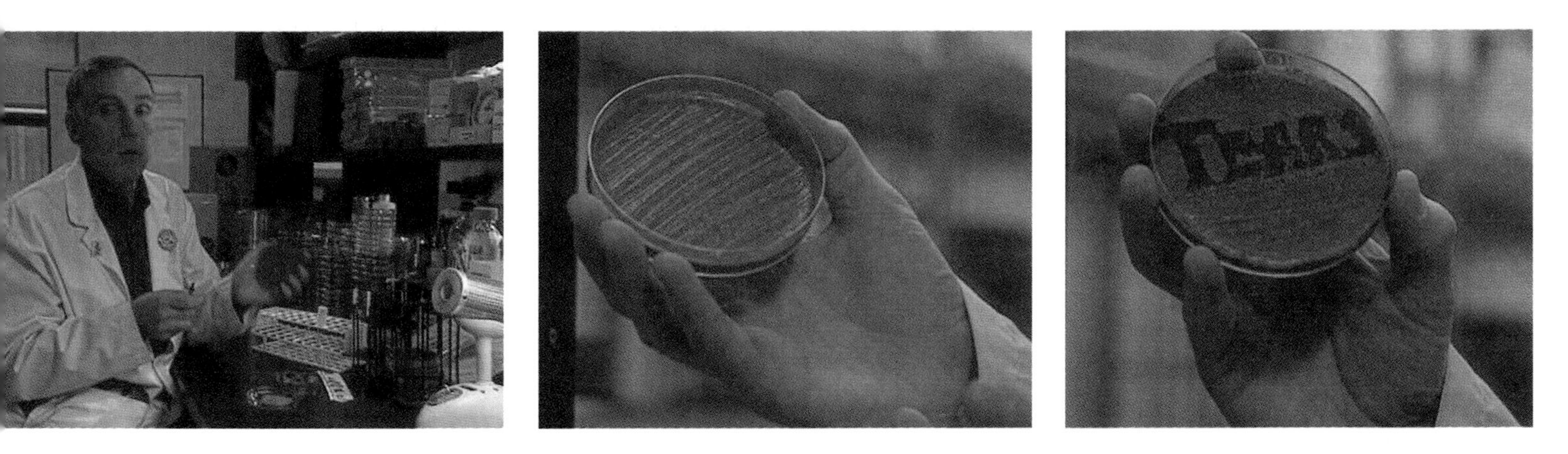

incubating overnight, the bacteria grew into a dense lawn of colonies, as you can see. Dr. Simor then collected some tears from a willing volunteer (well, slightly willing, and slightly paid) and used them to write the word "TEARS" on another lawn plate of bacteria. The next morning, hey presto! Wherever the tears had been dropped onto the plate, the bacteria had died. And as you can see in the photograph, the word TEARS is plainly visible as clear areas of uninhabited zones of agar.

The particular enzyme in lachrymal fluid (a medical word for "tears") that does the bug-killing trick has been identified and it is called *lysozyme.* What allows us to see clearly and keep our eyes free of infection—even when we have spit in them—is the antibacterial lysozyme combined with the rather efficient washer-wiper system of our eyelids and blink reflex. With those two defence mechanisms in action, we are able to keep the wildlife from establishing itself on our eyes most of the time. Even when we use our own tongue as a cheap lens bath.

Me and Someone Else's Big Mouth. The author rises to the occasion and the height of the Monell Center's icon.

4

The Worms' Turn

Mythologists have suggested that it wasn't a serpent in the Garden of Eden—it was the Primordial Worm, the mother and father of the progenitor of the wriggling and squirming ubiquitous multitudes to come. Perhaps it travelled with Adam and Eve in the enforced emigration of the Fall.

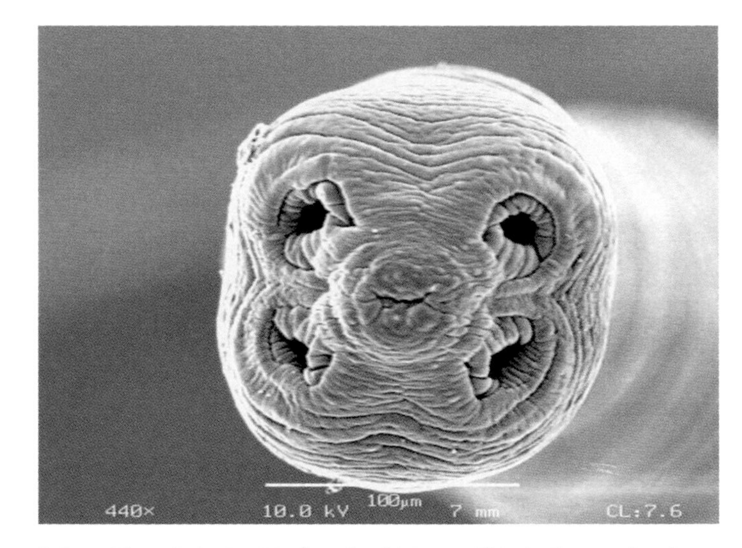

A Face That Only Its Mother Could Love. This is *Hymenolopsis,* a tapeworm that lives in rats. If you think it's smiling at you, turn the book upside down and it will scowl at you. Either way, it's very nasty and you shouldn't try to keep one as a pet.

Adam and Eve and Friends

Many mythologists believe that in the story of the Garden of Eden, the only other sentient resident was not a serpent—in the sense of a snake—but was actually meant to represent a worm. Whether they were actively involved in bringing about the Fall, people have believed for millennia that worms have played an extremely important role in the destiny of humankind. As indeed they have—and still do. Of course, it would be nice to blame the worms for all of humankind's misfortunes—especially since after we die we become (so it is said) food for them, thus giving them the last word in any argument. Nevertheless, there are so many different kinds of worms and the spectrum of their interactions with the human species is so incredibly broad, that it really is worth devoting some time to looking at them. And vice versa—well, not really, since we may be able to look at them, but worms don't have eyes and cannot look back at us. Which as you can see from the photos on this page may be no bad thing.

In this chapter, you'll see some of the astonishing ways in which worms affect humans—both very bad and (probably) very good. This may surprise you. At one extreme, of course, everybody knows about nasty worms—no intelligent person actually wants to acquire a tapeworm, for example (even if they are desperate to have their own pet). And tapeworms are by no means the most fearsome or the nastiest of the group. But the more you know about the bad ones, the more likely you can avoid crossing paths with them—and the more you know about the good ones, the less anxious you'll be if they are ever recommended to you as a form of treatment. (Yes, honestly.)

But first of all, some simple facts. There are basically four major groups of worms. There are the ones like earthworms (the *annelids*), then there are the tapeworms (the *cestodes,* which consist of long chains or segments strung together to make a flat ribbon), next are the flukes (*trematodes,* the ones in which the adults look like miniature vases), and most prevalent of all, the thin and rather featureless ones that are called

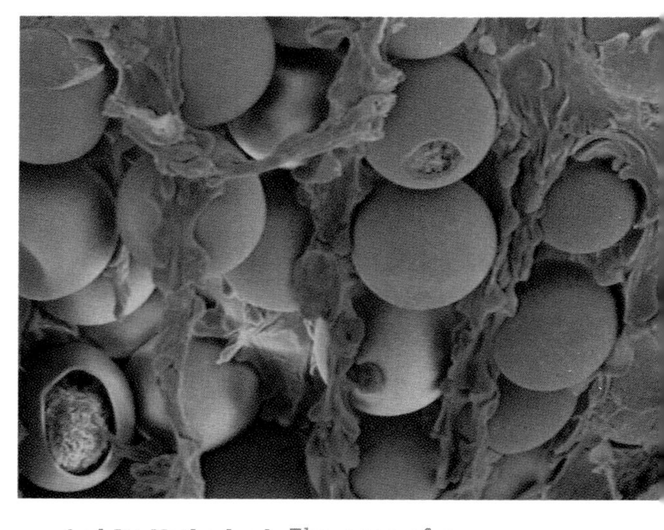

. . . And Its Motherload. The eggs of a *Hymenolopsis.* Apart from latching on to an intestinal wall, eggs are what a tapeworm does best. Not over easy, but easy to get.

nematodes (which include roundworms, hookworms, whipworms, and pinworms).

The nematodes (the roundworms and relatives) are an amazingly successful group and live as parasites in an enormous range of other species. In fact, it is often said that if you could magically make every animal group disappear except the roundworms, you'd still recognize the world. Cows, dogs, cats, rabbits, and almost every other animal would be recognizable by the collections of roundworms inside them—presumably they'd be walking around in their own characteristic but ghostly style that would help you work out what you were looking at. Humans would be recognizable by their roundworm populations in the colon, and you'd be able to make out the locations and outlines of villages and cities by the number of humanoid nematode colonies walking around. Presumably you'd be able to distinguish cinemas and football matches by the pattern of roundworms. It's not necessarily a very entertaining thought, but it's a sobering one, and it tells you a great deal about the genuine ubiquitousness of nematodes. They have got into everything. And they're still there. Or, rather, here.

Let's start with one of the slightly less threatening ones—a roundworm that in general prefers living inside pigs to living inside humans. Its name is *Trichinella* and for centuries, people have been puzzled about its mysterious *modus operandi.*

Tricky *Trichinella*

Spending time with Dr. Michael Sukhdeo in his laboratory in Rutgers University in New Jersey was genuinely a real joy. I don't think I have ever met a scientist more immersed in his subject and loving it. Michael and his wife, Suzanne—a co-researcher and very much a kindred spirit—together with the rest of their team create around them a wonderful atmosphere of creativity, knowledge, serious thinking, and fun. A relatively rare mixture in most research laboratories, in my experience.

Sukhdeo's credentials in the study of parasites are impressive, and I will give you just one small example. We humans can pick *Trichinella* up as an alternate host—that is, the kind of host that just wanders into the parasite's life cycle when neither of them really intended to meet. For humans, the problem starts when we eat pork that is infected with the larvae of the *Trichinella* roundworm. These larvae are wrapped up inside little cysts—neat and pretty rugged biological packaging. Curiously we are more likely to get infected when we barbecue pork that has been slaughtered and prepared locally (as opposed to inside large commercial organizations)—there have been several Oktoberfest festivals where sausages homemade from locally produced pork have caused this trouble.

Once inside a human's stomach, the *Trichinella* larvae come out of their little cysts and then move down the intestine until, at a certain point, they "decide" to thrash their way out of the bowel, breaking through the wall and setting off on a long and pretty complicated journey around the body. The big question was always this: What is it that tells a *Trichinella* to stop floating downstream and suddenly start its escape-through-the-wall-type of behaviour? Sukhdeo—like many of his

Bile Is the Solution. This man, Michael Sukhdeo, is the person who found out how bile steers the pork roundworm onto a new track.

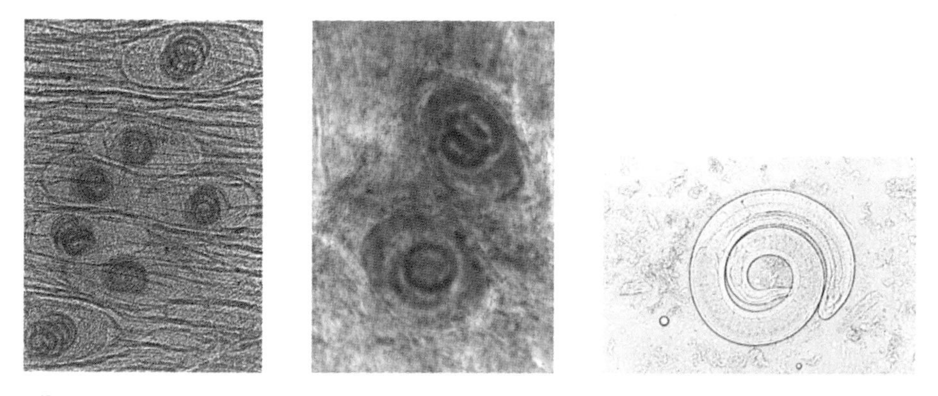

Bile creatures. These are *Trichinella* worms that have burrowed into the muscles of a pig and created the cysts in which they live. If the muscle they are living in is eaten, it is the bile of the carnivore that triggers them to break out of their cysts (bottom right) and begin the complicated journey through its new host.

predecessors in parasite research—tried all kinds of different triggers, and after many months was beginning to despair. However, unlike his colleagues and forbears, he was deeply imbued with the topic and an expert at forcing his brain to "think like a worm" (although, of course, worms don't think—or so we'd like to believe). Eventually he asked himself what changed in the worm's world as it made the journey from the stomach through the duodenum.

The answer was that at that point it met a stream of bile from the bile duct that joins the gut in the duodenum. So Sukhdeo took some newly hatched *Trichinella* worms and gently added bile to the solution in which they were sitting. To his amazement, they went crazy. Whereas they had all previously just sat there doing nothing, suddenly they started whipping about, curling and just as suddenly uncurling. Inside the small intestine, this behaviour pattern would immediately propel the worm up against—and then through—the bowel wall, where the journey upwards to the liver and then to the lungs would be a passive matter of following the stream. To me, that is the essence of really great biological research. The animals (or the cancer cells or whatever) are clearly doing something that is very specific and particular and serves some function, but we don't know what it is and why it happens. Finding the triggers that cause the animals or cells to do what they do is a stunning achievement. It does sound a little peculiar to say that a day spent in a laboratory with dead rats and writhing worms was exciting and inspiring—nevertheless, those are the two best adjectives to describe the Sukhdeos' unit.

The Zenith of Successful Parasitism

Trichinella is a fairly gentle introduction to the amazingly pervasive kingdom of worms, but although it is important, it is not the most successful or most widespread species parasitizing the human world. That title belongs to *Ascaris*.

If the *Ascaris* nematode were a human being, it would be Bill Gates. And I don't mean that unkindly. From an evolutionary point of view,

Ascaris has really made it big—it is present at all the major events in everyone's calendar and is intimate with everybody of almost every culture, gender and, indeed, species—at least until recently in the developed world.

In many ways, *Ascaris* is the planet's most successful parasite and exemplifies all the criteria of brilliant parasitism. It is widely distributed over the vertebrate world and has done particularly well in the human species—of the six and a half billion of us on this planet, it is estimated that at least one and a quarter billion have got one or more *Ascaris* worms living inside us. And, for the most part, they do not call attention to themselves. A good parasite likes it that way. A clever parasite does not want to upset its host too much, otherwise the host will not move around and reproduce and thus create more homes for the next generation of parasites. (In fact, from that point of view, something as lethal as the Ebola virus is a particularly bad parasite. As I mentioned at the beginning, that may be attributable to the fact that it has only crossed over to our species in the relatively recent past.) The one thing a parasite does not want to do is to bring about the death of its host quickly before the host has had a chance to mingle with other members of the same species and spread the parasite around.

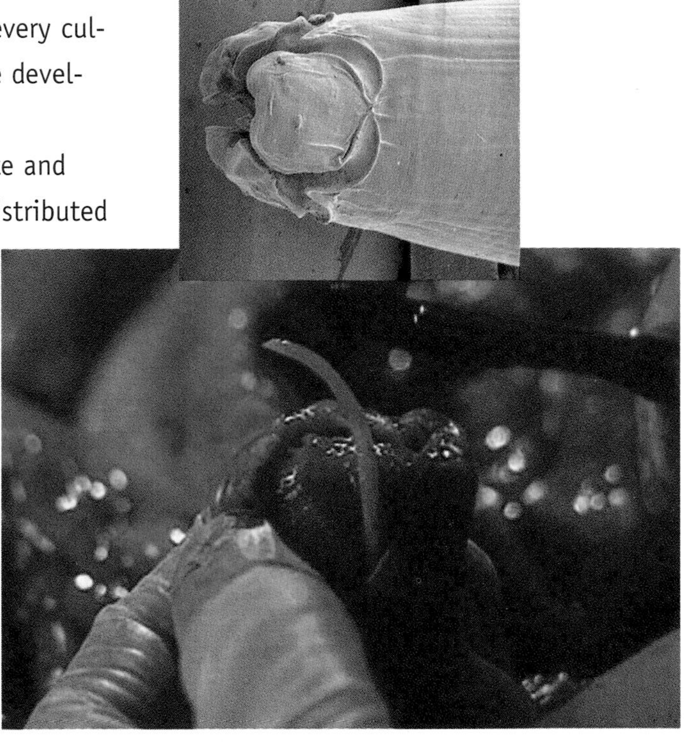

The Worm Turns (to face the camera). This is the unpleasant face of the most successful worm parasite of all time—*Ascaris*.

The *Ascaris* system works well and outside the body of its host species, its *modus operandi* is quite simple. The eggs of *Ascaris* get into the soil in the stool of other hosts. The soon-to-be host—humans, for example—ingests the eggs either directly (such as when children play in the dirt and put their hands in their mouths) or indirectly (such as when fruit or vegetables are contaminated with feces containing Ascaris eggs and are eaten without proper washing).

Once inside the host, however, the *Ascaris* system suddenly becomes pretty complicated and intricate—the worm literally does a double-circuit of the human body as it goes through its life cycle. First of all, the eggs

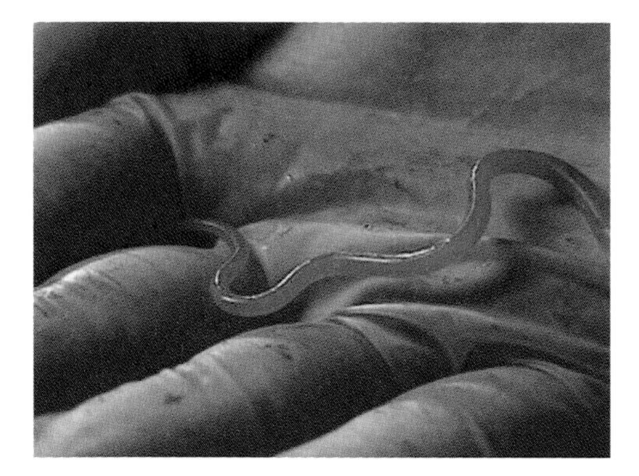

Trouble in Hand. These are relatively small *Ascaris* worms—really big ones can grow to two or three feet or even more.

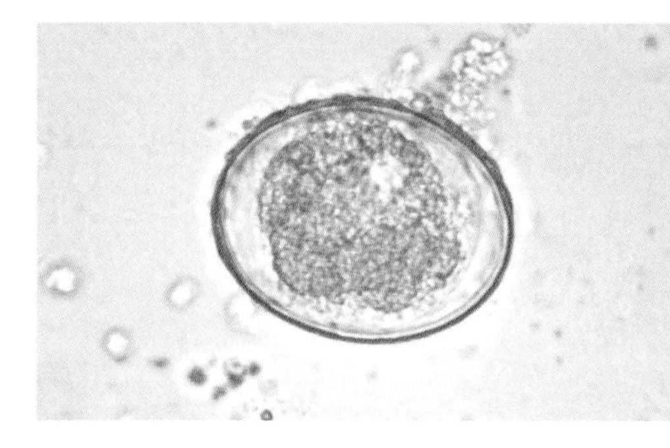

Ascaris eggs (above and below)

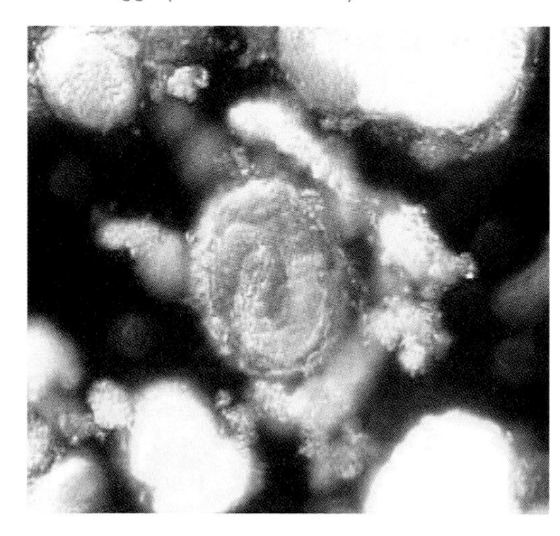

hatch in the acidic environment of the stomach and are moved on into the small intestine. There they migrate through the wall of the intestine and travel to the lungs—in a pattern of mysterious but purposeful exodus that makes the Monarch butterfly's migration look as complex as the wanderings of a toddler in a playpen. Inside the lungs, the young *Ascaris* larvae turn out to be a very mild irritant to the host's airways. Appropriately, from the host's point of view, the irritation sets up a cough reflex. This allows the host to cough up the irritation, but at the same time, it gives the *Ascaris* a free ride. The larvae are coughed up into the throat and then immediately swallowed; now—being considerably more mature than they were on their first visit when they were just eggs—they journey down the entire gut and eventually settle in the colon as adult worms. There they do what they do best—almost nothing. They hook onto the wall of the colon and bask in a food supply that was yours originally, but is now shared.

In the great majority of cases, one or two *Ascaris* worms inside you would be virtually unnoticeable. In fact, it is quite common for people to be unaware of the infestation until they see an adult worm passed in their feces. Adult worms can occasionally be coughed up as they pass through the lungs, but this is very rare. If the infection happens to be very severe, the person may suffer from anemia or even malnutrition, but again, these problems are the exception rather than the rule.

Ascaris is, as I have said, a good example of an astonishingly successful parasite—we scarcely notice it's there. We do not know for certain—although it is probably the case—that this kind of very effective parasitism, verging on commensalism, has emerged through millennia of co-evolution. As we evolved together, we each did a little better when we developed some form of tolerance. This may be the future for many of the parasitic infections that are troublesome to us now—it is possible that some of the horror stories you will see in this chapter may indeed evolve into tolerance many thousands of years from now, In the meantime, however, we have to distinguish between the "really nasty worms" and the "not so bad really" worms. In this context, *Ascaris* is definitely one of the latter.

Chance Encounters and Other Flukes

Even in paradise, there were serpents, and the tranquil Lakes that abound in Quebec are—by any standards—frontrunners in the paradisiacal contest. About an hour's drive north of Mont Tremblant, a well-known and much-loved resort in Quebec, a wonderful and friendly community of Korean families have wisely built their cottages and holiday homes on the shores of one such lake.

In 1993, Dr. Francis Han decided to have a special party for his family, a few of his best friends, and some visiting Korean dignitaries. The most favoured component of the meal was a particular sushi dish of which the main constituent was a fish known as the white sucker. As Dr. Han explained to me, the preparation of the dish was an important part of the occasion, and the fact that the fish had been caught less than a hundred yards from the patio on which it was served simply emphasized the beauty and harmony of the area and of the occasion. All was as it should be.

Unlucky for Some—Flukes. Flukes of various shapes and sizes that parasitize different species.

This made it all the more upsetting when, soon after eating the delicacy, nineteen of the guests suddenly developed severe abdominal pain, fever, and vomiting, and required hospitalization. The stricken guests were all taken to the Centre for Tropical Diseases in Montreal, where Dr. J.D. McLean recognized the problem.

The culprit turned out to be a worm, *Metorchis conjunctis,* a type of liver fluke, which, as it happens, is a known inhabitant of the freshwater white sucker, but not of the marine white sucker. Like many of its group, *Metorchis conjunctis* has a complex and very upsetting life cycle.

The worm hatches in the stomach and then swims up the bile duct—perhaps following the chemical signal up the gradient as the "smell" of bile gets stronger. (By contrast, the larvae of the *Trichinella* tapeworm, as Michael Sukhdeo has so ably demonstrated, do not follow the trail of bile, but respond to it by thrashing around and thereby penetrating the wall of the bowel.) The *Metorchis* crawls along in the same way as an inchworm does: it has a sucker at each end and arches itself up and moves forward by moving the front sucker. As it progresses into the liver, it starts making the poor patient—the Han family in this case—feel suddenly very sick indeed.

Metorchis is not a common cause of abdominal problems, but fortunately the diagnosis was made quickly in the Han outbreak, medication was administered soon after, and all of the victims have made a complete recovery.

Not the Usual Problem. Dr. J.D. McLean realized that the members of the Han party who came down with severe abdominal pain and fever were not suffering from a simple or common cause of food poisoning.

When I talked with Dr. Han and some members of his family and they looked back on the incident, they made the important point that nobody could have expected that much trouble in Paradise. The lake is absolutely lovely, the shoreline and the woods around it are stunning, and it is easy to imagine that catching some special fish would have just added to the anticipation of a great party. The white sucker is known for its soft and delicate-tasting flesh, Dr. Han explained, and nobody would have suspected that something caught a few hundred feet from the cottage in those peaceful and tranquil surroundings could contain anything sinister. Perhaps that is an important—although unfair—fact of the wildlife on this planet—even the Garden of Eden had serpents.

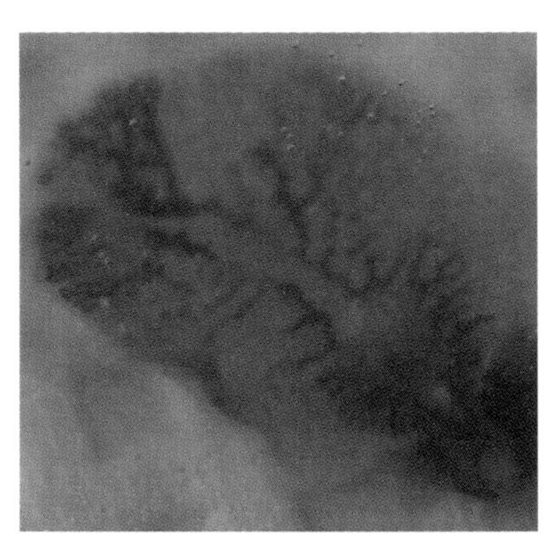

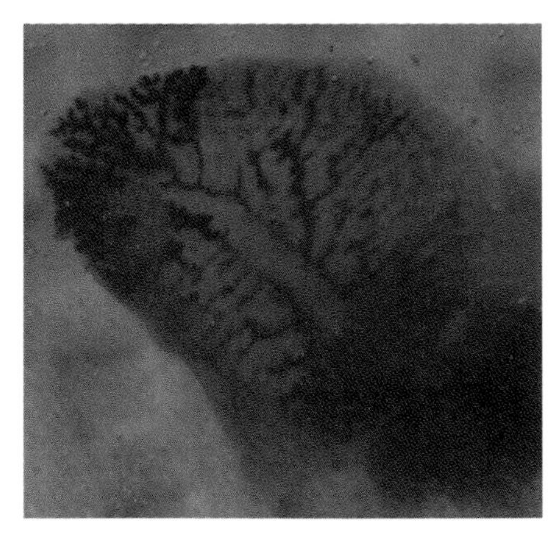

The Culprit from the Lake. This is *Metorchis conjunctis,* a type of liver fluke that colonizes a bottom-feeding fish called the white sucker and caused the serious problems at the Han family party.

Schistosomes: They're Everywhere and They're Hated

Another fluke (or trematode)—*Schistosoma*—is a real health problem for us on a worldwide scale. It has a complex and highly efficient life cycle and its three most important species (*mansoni, haematobium,* and *japonica*) wreak widespread havoc on a massive scale in Egypt and most parts of Africa, in the Philippines, in Japan, China and Indochina, on the Arabian peninsula, in the Middle East, and in South America. The reason why this particular fluke is not indigenous to North America is that its life cycle depends on a freshwater snail and, as it happens, that particular snail doesn't live on the North American continent.

Like most flukes—including *Metorchis conjunctis,* the liver fluke that the Han family encountered in freshwater white suckers—*Schistosoma* has an intricate life cycle consisting of several stages and depends on a minimum of two types of host (a snail and a mammal) to complete that cycle.

The larval stage of the fluke get inside us by breaking through the skin. They are very small, but their entry usually produces an itch after the event. In fact, *Schistosoma* is one cause of swimmer's itch in certain areas of the world. What happens after the larvae have broken through the skin depends on the species, but usually *Schistosoma* ends up living in, and doing great damage to, the bladder (*S. haematobium*—the cause of bilharzia) or the liver (*S. mansoni*).

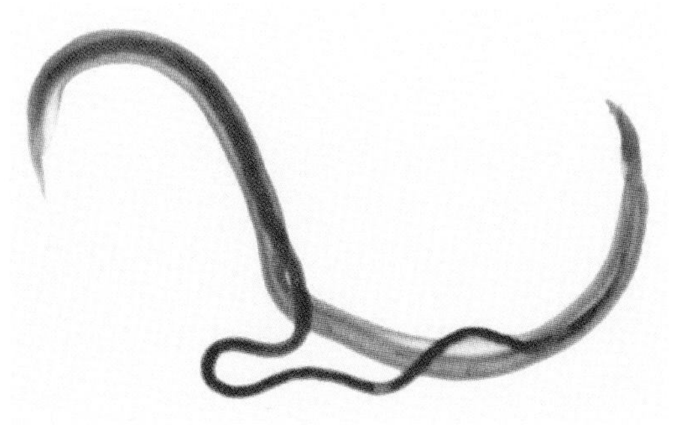

In Flagrante Delicto. The male *Schistosoma* (actually the larger one on the outside) and the smaller female.

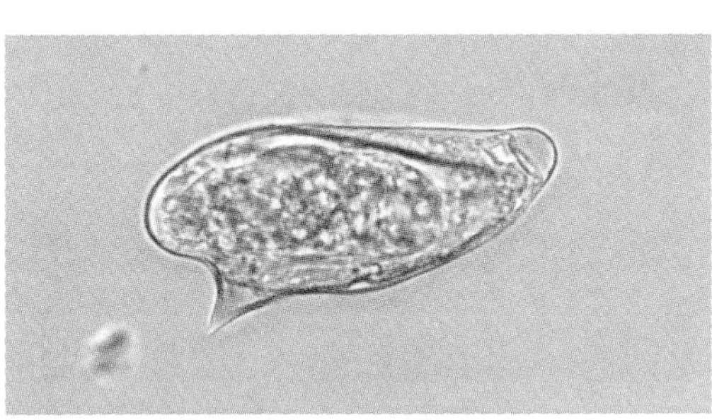

A *Schistosoma's* egg.

Snails' Pace and a Snails' Place. In China major public-health campaigns have been directed at freshwater snails. Nowadays they are more commonly found in the market—as a menu item—instead of in the rivers as an intermediate host for worms.

In some parts of the world—particularly in China—major public health campaigns have been directed against freshwater snails precisely to break the *Schistosoma* life cycle and reduce human infestations; these have achieved a great deal. Other public-health measures, such as making sure that people do not urinate into rivers (because the urine may contain *Schistosoma* eggs and allow infection of snails), have also been effective.

As is common with the nastier types of human wildlife, we may not be able to exterminate the species (although I am sure we would like to), but the more we know about the details of the organism, the more we can take intelligent precautions against being infected by it.

By Hook and by Crook

In Michael Sukhdeo's lab, the little pebbles of charcoal in one particular Petri dish seemed to be twinkling. On closer inspection, I could see that it wasn't the charcoal itself that was twinkling, it was what was on top—the movement that I discerned actually looked as if hundreds and hundreds of tiny little people were standing on the charcoal waving like shipwrecked tourists at an airplane from their desert islands. Only these weren't civilized shipwreckees, they were worms—and they weren't waving in order to attract anybody's attention. They were waving about so that if any mammalian body happened to sit down nearby or walk past them, they could hitch a ride and burrow into the poor victim's skin.

Paradise (Caution: May Contain Serpents). To us a sandy beach like this offers the perfect holiday environment—to the hookworm it's a great place to meet us and bore into our skin.

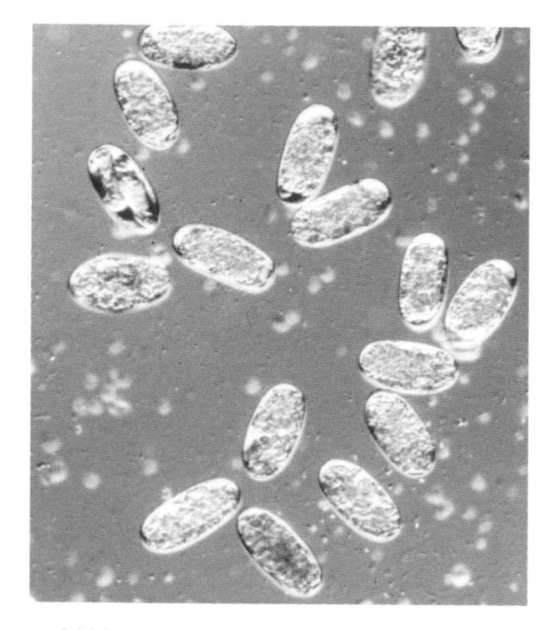

Forbidden Fruit. Looking like a view of some exotic vineyard or olive grove, these are actually the eggs of a hookworm.

That is why this type of roundworms are called "hookworms." Although they are small, they are very savage. Or to put it another way, although they are very savage, they are small. Either way, the combination of those characteristics has made them quite successful as parasites and a formidable foe.

Their size and their ability to penetrate human skin make hookworms very difficult to detect as they enter through the skin of your feet or thighs. Often, as they go, our bodies respond to their passage by creating inflammation around the track. This inflammation can sometimes be seen from the outside—and is then known as *cutaneous larva migrans.*

As you can see from the picture at the bottom of the facing page, the hookworm's path is tortuous. The worm makes its way, eventually, to your intestines and is not harmless. Unlike most intestinal worms, they take a fairly big bite of the bowel wall and simply hang on. That bite breaks into blood vessels and provides the hookworms with a ready source of their staple diet—blood. If a victim happens to have a heavy infestation of hookworms, he or she can become quite anemic from loss of blood, and on a worldwide scale this is a significant health problem. Hookworms are wildlife that we need to know about so that we can try to ensure that our paths don't cross.

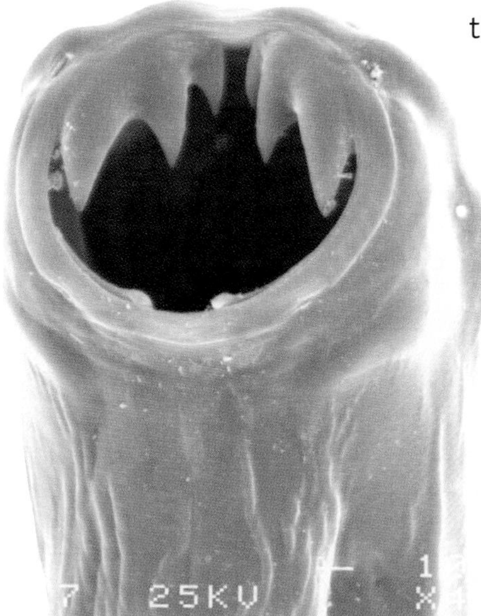

The Sucker That Is Born Every Minute. The business end of the hookworm, perfectly adapted for fastening onto the wall of the host's intestine.

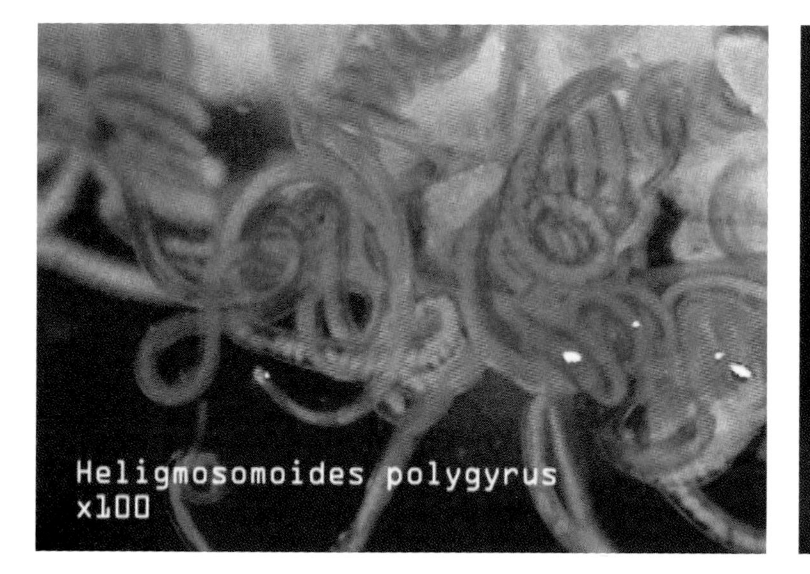

Boring Worms. These hookworms are programmed to wave about until they encounter a piece of mammalian skin and bore straight through it.

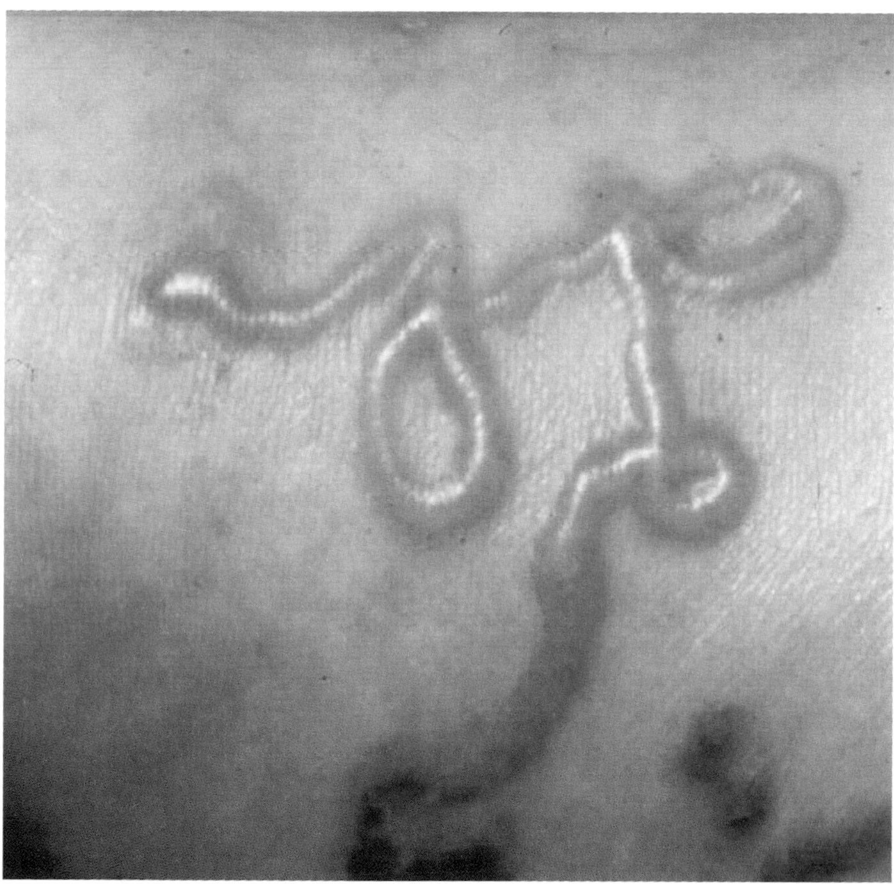

He Went Thataway. The track of larva under the skin (*cutaneous larva migrans*) may cause irritation and inflammation, thus showing its circuitous path.

Tapeworms—The Train We Would All Like to Miss

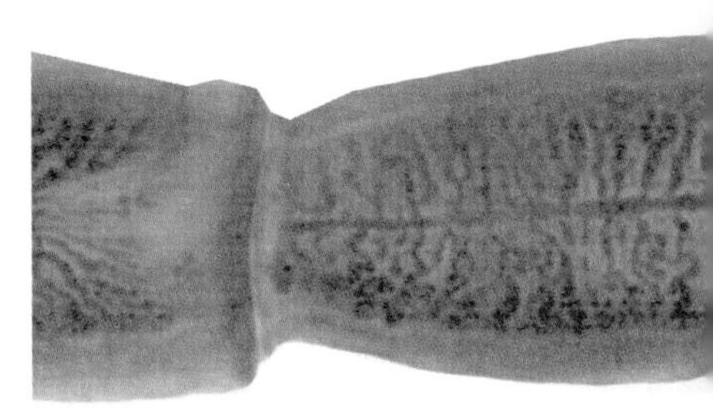

Every roundworm—and every hookworm—is a single animal, complete in itself. Tapeworms, on the other hand, are different—they are segmented and, to some extent, each segment of the tape is a self-contained unit. This is actually quite a difficult concept to grasp, but it may help to think of a roundworm (or a hookworm and even an earthworm) as being a bit like a very long, very thin bus or truck. The bus—even if it were a mile long and three feet wide—would still be a single vehicle, and if you cut it in half, it wouldn't run. Tapeworms, in some respects, are more like railway trains—just as trains are collections of carriages or wagons strung together, so tapeworms are

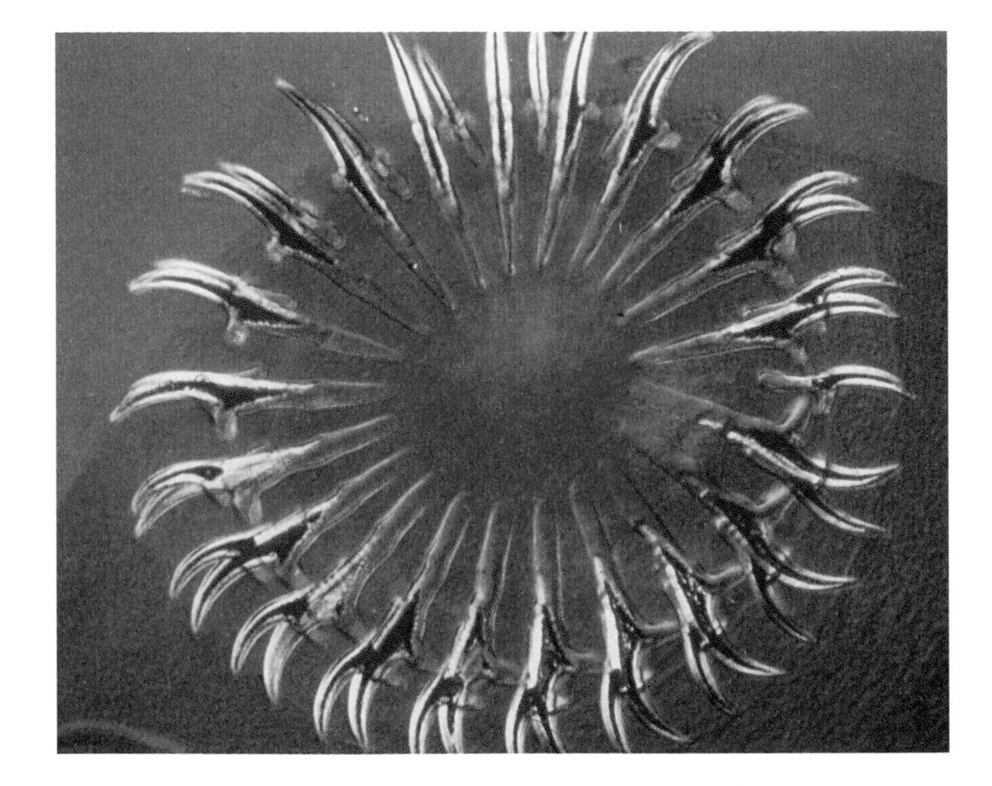

Not a Hollywood Musical Seen from the Air. This is not a still from a Busby Berkeley musical—it's actually the head of a tapeworm, and what you are seeing is a circle of hooks that will allow the head (*scolex*) to fasten itself to the host's bowel and start making little baby segments.

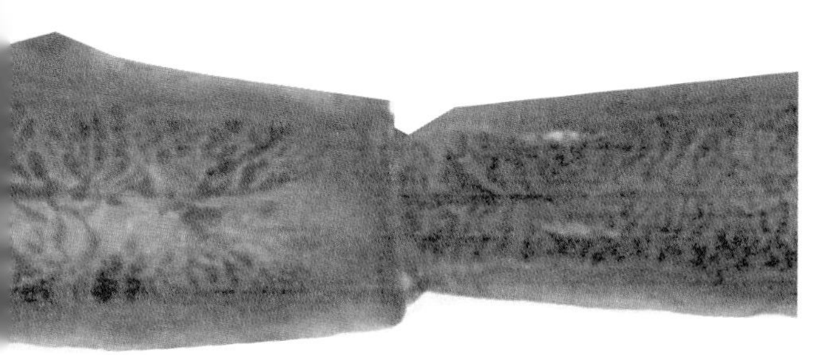

a long collection of units or segments (for which the correct scientific term is *proglottids*). Unlike trains, however, in tapeworms the engine actually creates the carriages rather than simply pulling them along. It works like this: the head of the tapeworm, called the *scolex,* fastens onto its host's bowel and grows by making daughter proglottids that, at first, remain connected to the head end. So the chain forms, as it were, from the front end, with each daughter proglottid moving the existing siblings down a notch.

What is even more unusual is that as the proglottids form from the scolex end, they are initially genderless (they are neither male nor female). In some species of tapeworms, though, the proglottids mature first into males and then into females. So the proglottids half-way down the length of the worm may show characteristics of male gender, and some inches farther down the chain, the same proglottids will become female. In many species, the newly matured females (or "used-to-be-males") are fertilized by the younger male segments—an example of true hermaphroditic self-fertilization—and become pregnant ("gravid") with eggs, at which point they usually separate from the worm chain, fall out of the host in the feces, and are ready to start their own families in new hosts.

Tapeworms are perhaps mid-way in the spectrum of horridness. They can cause malnutrition and anemia if the infestation is heavy, and they are certainly to be taken seriously. On the worldwide scale of human suffering, though, they are certainly well below the next member of the rogues' gallery of worms that causes untold pain and suffering.

Not Quite a Conga Line. Each *Proglottid,* or section of the tapeworm, by the time it is ready to detach itself from the worm, is filled to the bursting point with eggs.

Unmitigated Horror—The Guinea Worm

The medieval painter Hieronymous Bosch gave the world many unforgettable graphic visions of hell, with poor condemned souls being tortured in every imaginable way at the hands of gleeful, sadistic, pitchfork-wielding demons. I don't know if Bosch had ever heard of the guinea worm, but if he had he might have portrayed it as an instrument of torture even worse than fire or brimstone or the combined effects of both.

From the human point of view, the guinea worm (*Dracunculus mediensis*), also known as the fiery serpent, is pure, unmitigated evil and causes continuous, excruciating pain and suffering. Its life cycle is ingenious and has a certain terrible perfection in its design—just like a finely engineered instrument of torture. It starts as a small larva in contaminated water, usually in poorer areas of Africa. The larvae are eaten by tiny little water fleas, which ferry them around, helping them to distribute themselves over a much wider area than they could reach under their own steam.

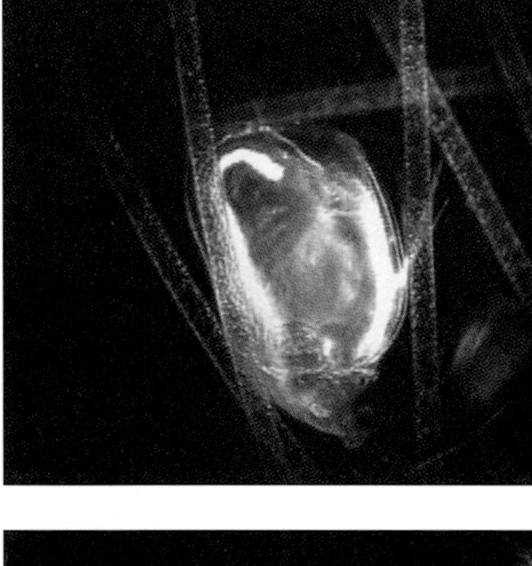

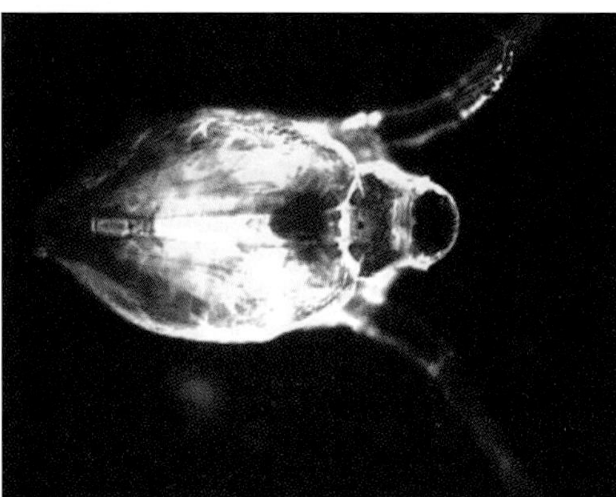

The Water Taxi. This little water flea (a *Copepod* to give its correct name) is intermediate host to the vile Guinea worm.

For humans, the problems begin when we drink water contaminated with water fleas that have Guinea worm larvae in them. Inside the human stomach, the acid kills the water fleas and releases the Guinea worm larvae—which are resistant to our gastric acid. They move along into the small intestine and then begin a dogged and cruel migration. They bore their way out of the intestine through the intestinal wall, and steadily make their way to the subcutaneous tissues of the leg. Precisely how they evolved this extraordinary pattern of migration, and what cues they use to mark their route, is still largely unknown. But they do it—and it takes just under a year. Along the way, they mate. The female is somewhat larger than the male and after they have mated, the male dies and is actually absorbed by the female. (It sounds disgusting—but it conserves more natural resources than a divorce.)

At this point, the adult female, which by now may be one to two feet in length, steadily moves towards the skin (usually on the legs although it can emerge from other areas). As it reaches the surface, it creates a blister in the skin that causes intense burning pain. The poor victim will

The True Face of Disability. In modern paved cities, a crippled or disabled person can get around quite easily—but if this is the only bridge available, a painful disabling condition such as a guinea worm infestation can imprison its victim.

try to relieve the burning sensation by plunging his or her leg into cold water—usually the river. The change in temperature causes the skin over the blister to break down, and the adult worm is exposed, allowing it to release millions of young larvae into the water. Which is when they get eaten by a passing water flea, and the whole cycle begins anew. The entire cycle is—unfortunately—brilliantly successful because, at each stage, the worm has evolved strategies for changing the behaviour of its host (relieving the painful leg by immersion) or using the features of the host to its advantage (the mobility of the water flea).

As regards treatment, the story gets still worse. One might be tempted to grasp hold of the protruding end of the wriggling worm, grit one's teeth and just pull. This would be—literally—fatal. If the body of the guinea worm breaks, with half of its body still under the skin, it will die

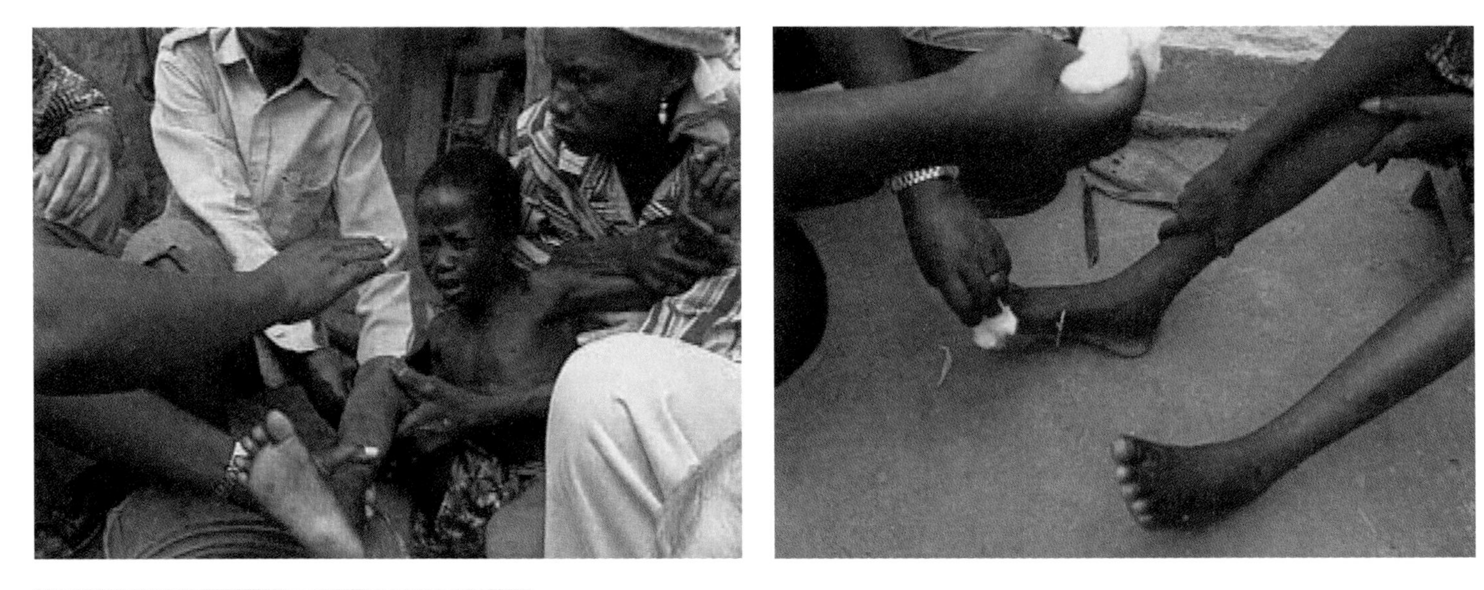

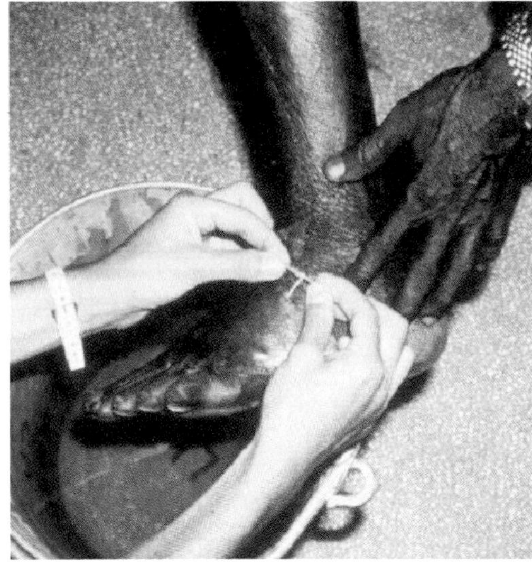

and as it dies it releases large quantities of proteins that will precipitate a sudden and lethal shock reaction. In other words, any attempt at a quick cure will kill the poor victim. Somehow, this sad fact has been known in almost every culture for which we have records, and from earliest times the treatment has involved slow and daily prying out of the guinea worm, inch by inch. The incredibly painful and slow process is usually done with a sliver of wood roughly the size and shape of a pencil. This treatment of guinea worm infestation is so embedded in human culture that many historians believe that the symbol of medicine and healing—the caduceus—is actually a representation of a guinea worm being wound onto a stick, as opposed to the traditional view that the symbol represents the magic healing staff of the god Apollo, given as a gift to the world's first doctor, Aesculapius.

Whether the guinea worm and the details of its treatment were known to the ancients, there is no doubt that it need not be a permanent feature of our species. Eradication of the worm as a human parasite can be achieved relatively easily and, although there is still much work to do, current procedures have already reduced the incidence of guinea worm infections dramatically. The key is to ensure that human drinking water

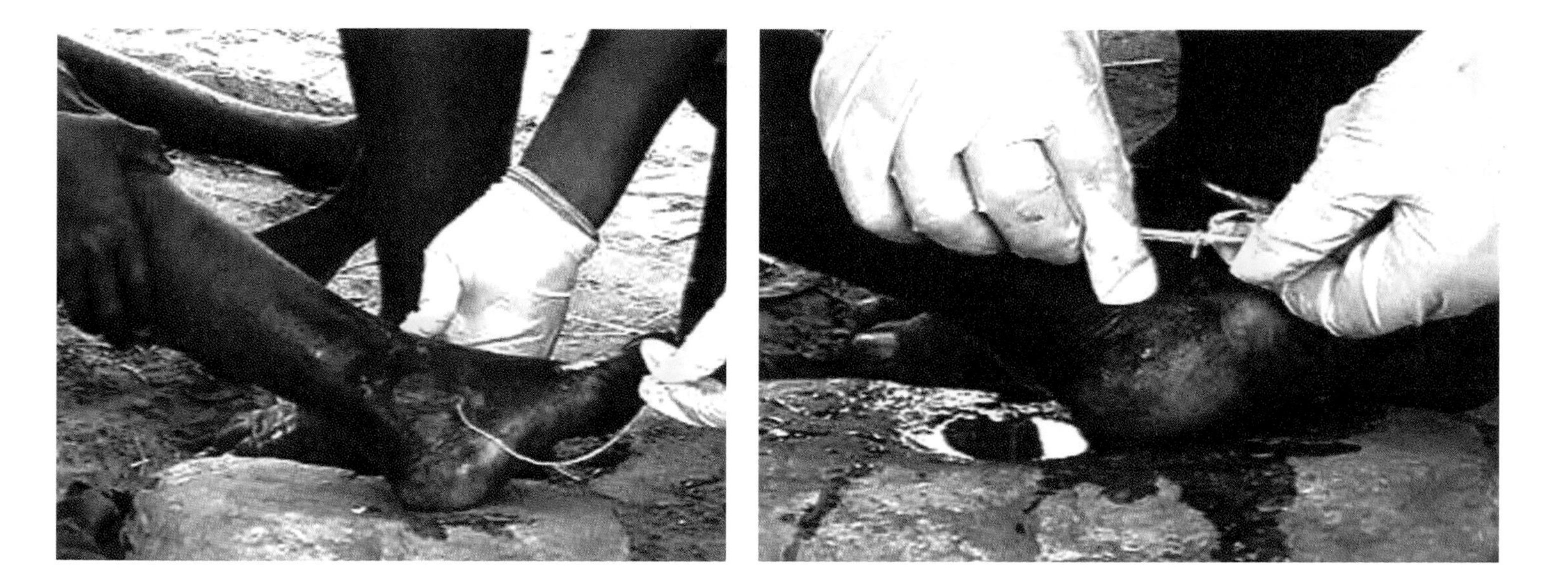

is not contaminated by water fleas—and that is actually easier than it sounds. Filtration of drinking water through filters as simple as a few layers of cloth effectively removes all of the water fleas. When this is done in conjunction with educating local residents to never drink unfiltered water, the guinea worm could one day be nothing more than a minor footnote in history and mythology, a mysterious symbol that has no association with the pain and suffering we see today.

The Healing Touch. The caduceus is the internationally recognized symbol of medicine and healing. Could it be a representation of Guinea worms, *Firey Serpents* as they are known, being removed. Take a look over the images on these pages and decide for yourself.

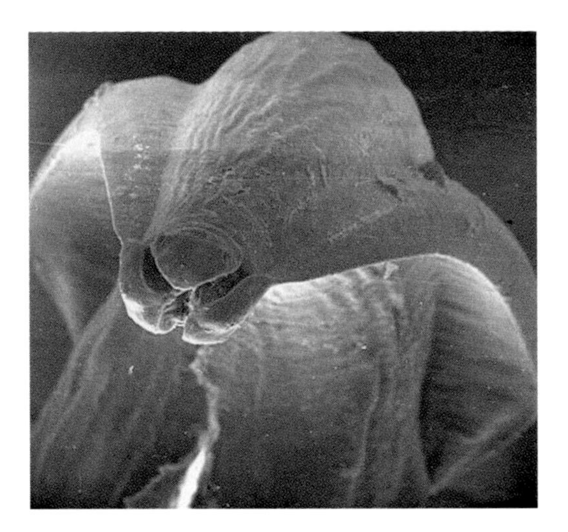

Puppy Love. This is the *Toxocara* worm, which inhabits about 80 per cent of puppies under the age of six months.

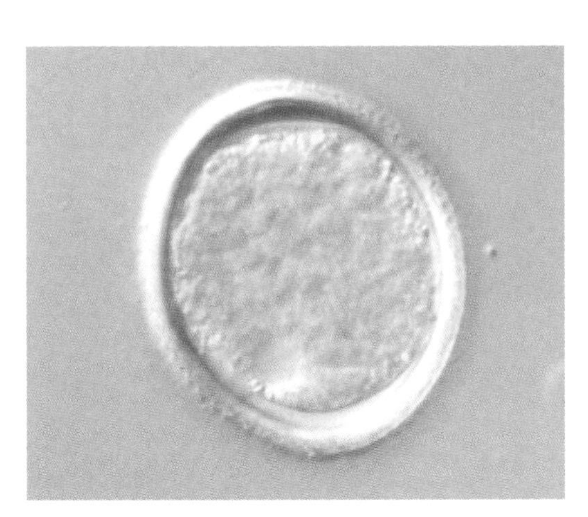

Travelling Light. *Toxocara* in its travelling form.

Toxocara—An Accidental Tourist on Planet Human

It is most unfortunate that the paths of humans and the worm called *Toxocara* cross—the meeting doesn't really help either party. *Toxocara* lives inside dogs and is the most common parasite in young puppies. It doesn't cause very much trouble for the puppies; it's only when we humans inadvertently step into its path—and when the *Toxocara* tries to remedy that fault—that the problems begin.

What you see in this photographs on the next page are the cyst stages of the *Toxocara* worm—an egg-like structure with a worm inside waiting to hatch. The problem is that cysts like this in the stool of

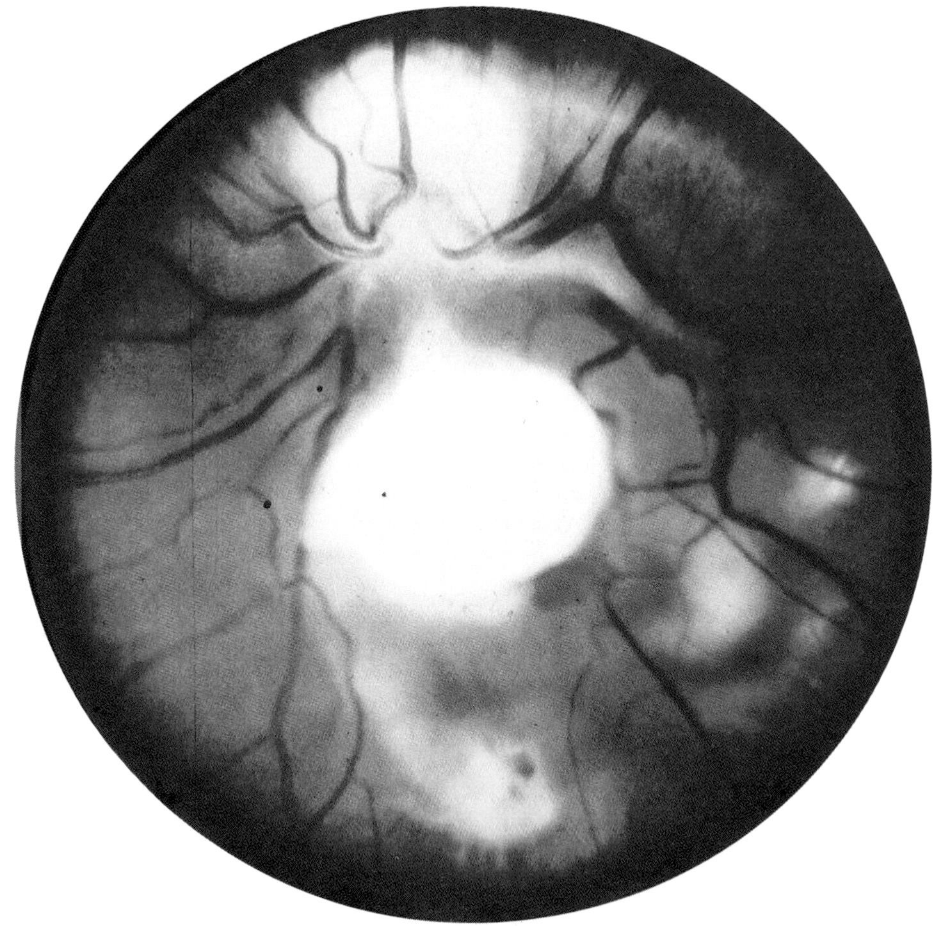

The Wrong Place, the Wrong Time. This is what the worm *Toxocara* looks like as it blunders through the human body—here ending up on the retina at the back of the eye—in its attempt to make its way into another more suitable host, such as a dog.

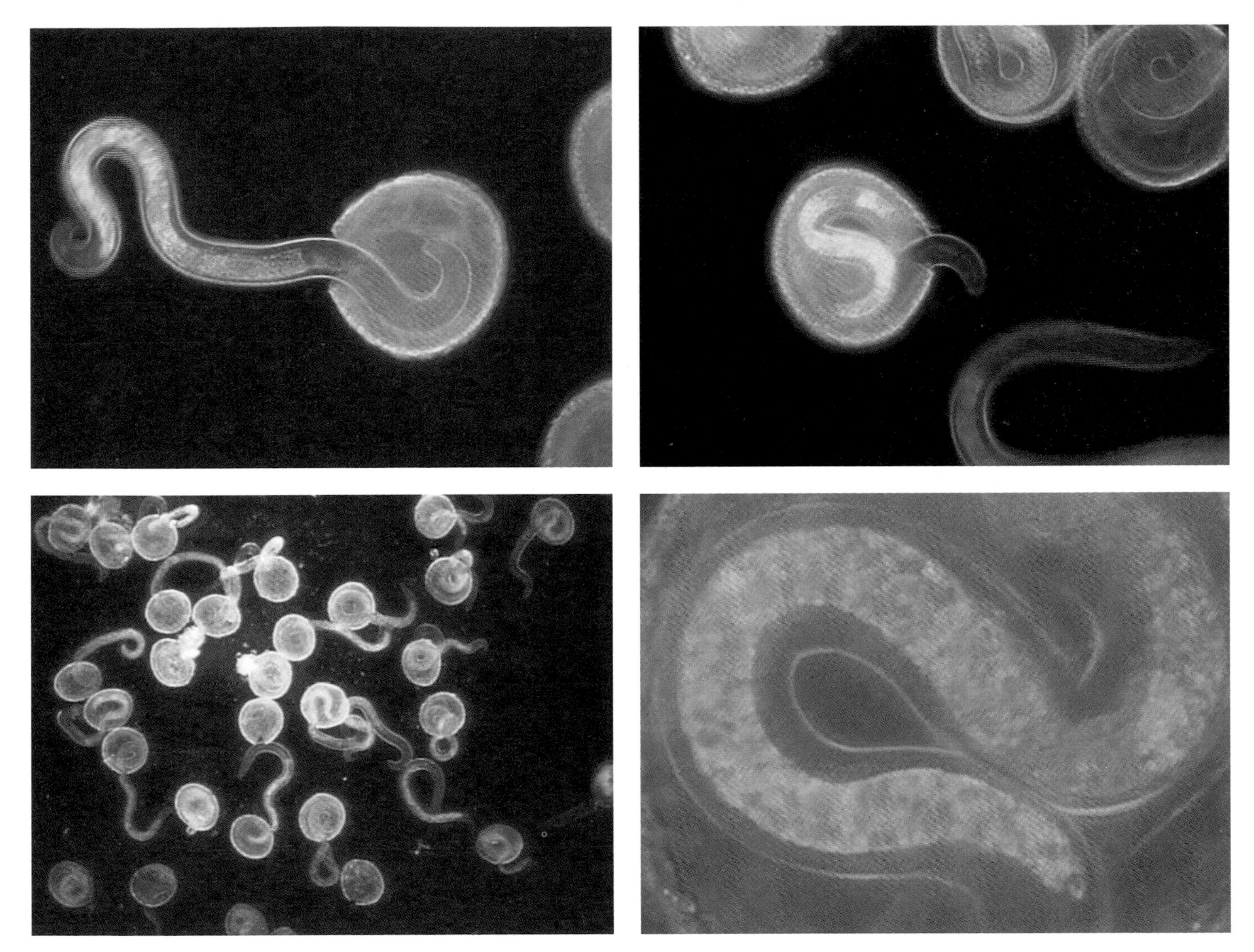

Unpacking. The unpleasant *Toxocara* worm now in its adult stage emerges from the security of its cyst form.

domesticated dogs can quite easily make their way into humans. Dog feces—as we all know—really travel. They get underfoot when we walk in the street, are readily transferred inside on the soles of our shoes, and end up on—and then in—the carpet. Children playing can easily transfer the cysts. Once inside a human, however, the *Toxocara* doesn't really like its environment and is "programmed to find its way out." The trouble is

Amazing Facts That You Can Tell People Round the Cooler

1. The staff-with-two-serpents-coiling-round-it symbol that most medical associations use as an emblem is called a *caduceus*.
2. The caduceus is traditionally believed to represent the staff of the god Apollo with two magical snakes—but it might actually represent the horrible guinea worm being wound out of a person with a stick.

that in doing so it causes intense damage. As Dr. Kerr, the veterinary surgeon who, incidentally, lent us Buddy, the fluffy dog in the picture on page 28, explains, the *Toxocara* worm doesn't really "want" to be inside a human, it wants to be inside a dog. So it starts to blunder around inside the human, trying to get out. In its blunderings, it can emerge through the eye (although, thankfully, that is fairly rare) or even the brain or heart. The damage occurs not only as it makes its path, but also when as the human body reacts to the worm, causing inflammation.

In its place—which is basically in the dog!—*Toxocara* isn't necessarily a very severe problem. About twenty percent of adult dogs have got it—but it is very much commoner in young puppies. In fact 80 per cent—that's four out of every five—puppies below the age of six months have got it. Because the *Toxocara* worm can cause such trouble—albeit in a few cases—when the eggs pass into the dog's feces and thence to new victims, deworming is something every dog-owner should take seriously.

Bridgend—A New Beginning

What you're going to read and see here may—at first—disgust you and give you an attack of the queasy heaves. But as you begin to see what's going on, I think you'll find it extraordinary and actually rather inspiring. At the Princess of Wales Hospital in Bridgend, Wales, a group led by Dr. Steve Thomas, the director of the Surgical Materials Testing Laboratory, is pioneering a method of treating severe and treatment-resistant infections by using a biological therapy that is both very old and astonishingly new.

The biological agent the team is using consists of maggots (and I'm sorry if that makes you queasy). In fact, these particular maggots—which

Ugly But on Our Side. These are adults of the greenbottle fly (*Lucilia sericata*) as we might see them buzzing around the garbage—in their young maggot form, however, they can be useful tools for healing.

look to the casual observer like a kind of worm (as you can see from the photographs above)—are the larvae of the common greenbottle fly. The greenbottle fly, *Lucilia sericata,* is the rather big and ugly glossy thing that you see from time to time buzzing around your house. They quite like spending their time in your garbage or on decomposing and putrefying things. They may or may not be of blameless character in themselves, but they certainly keep bad company and hang around low-life districts and environments.

Nonetheless, it has been known for about a hundred and fifty years that maggots—apparent metaphors of rot and putrefaction—are not merely signs of bad news and imminent dissolution. In fact, during the Napoleonic Wars, it was noticed that injured soldiers who had maggots in their wounds actually fared better than those who did not. Somehow, the maggots were keeping the wounds relatively clean. That would have remained just an interesting and inexplicable item of military intelligence, had not some researchers in the 1930s taken it seriously and put forward the revolutionary idea that chronic wound infections could be treated with maggots. The research was just getting under way when along came antibiotics—and with them the promise of a cure for all major infections.

Fifty years later, as we all know, it turns out that the optimism that we were at the dawn of a chemical utopia was a tad premature. Some organisms have never been sensitive to medications (we don't have pharmaceutical agents to treat most viral infections, for example) and some bacteria that were reliably sensitive to antibiotics have developed resistance to them. In some cases, the infection is situated too far away from the blood supply, so that no matter how much antibiotic the patient takes, the medication doesn't get to the bacteria in a high enough concentration.

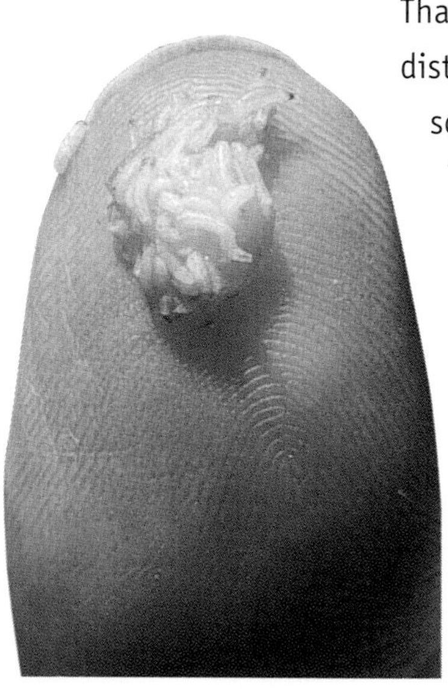

The World's Smallest Surgeons Meeting and Conferring About Their Next Case. Always ready to help.

That last problem—an infection that is too distant from a blood supply—is what can sometimes happen in bad ulcers of the foot. These can develop in a person who has diabetes, a condition that in itself affects the blood supply by narrowing the ordinary small arteries and capillaries. Once the bacteria have established themselves in an ulcer in a diabetic person's foot, for example, the infection can be very very difficult to treat indeed. The increased sugar levels due to the diabetes make it easy for the bacteria

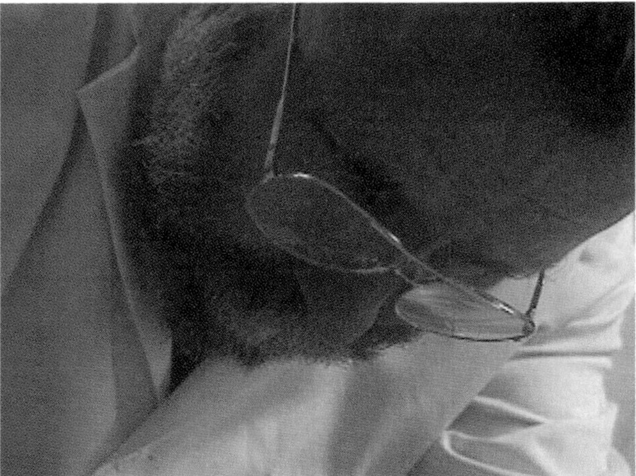

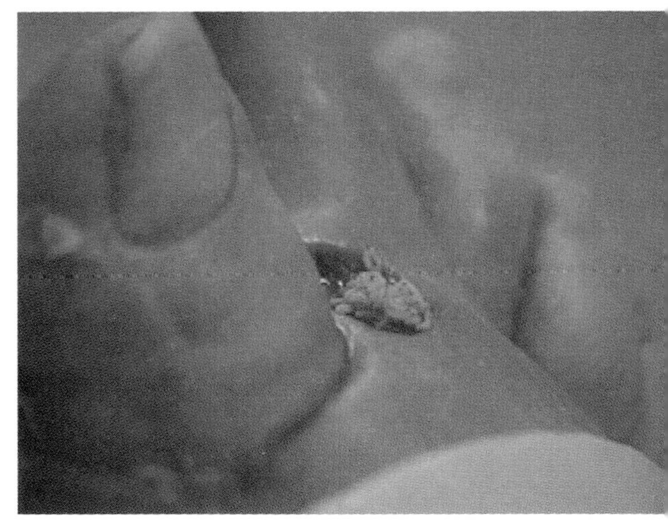

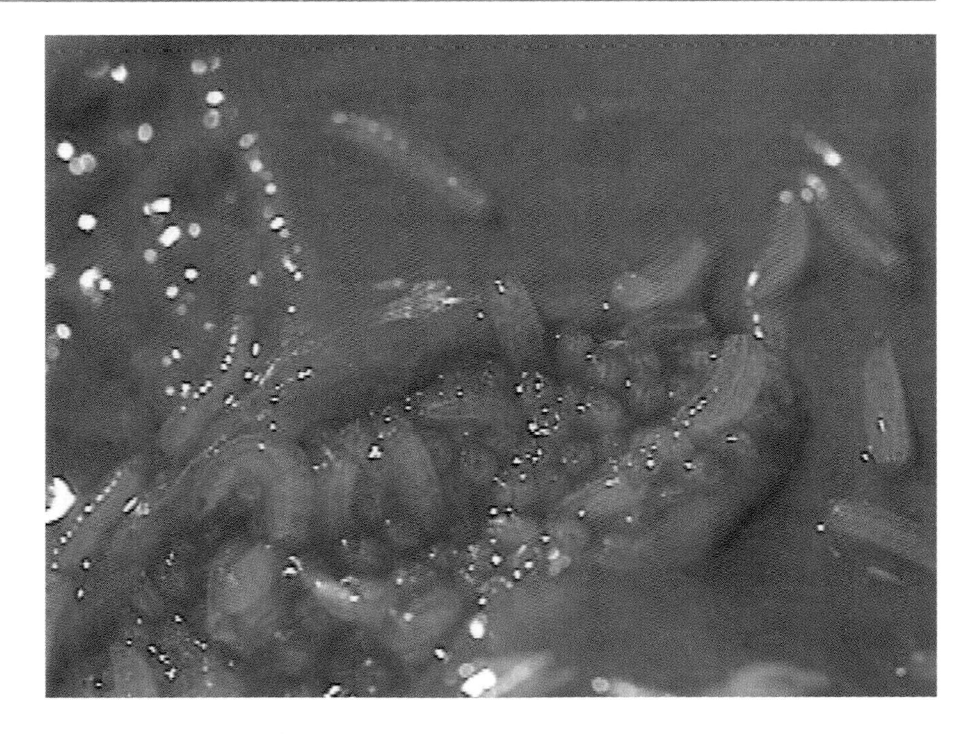

Not Easy to Look at. These pictures probably make you feel queasy, but actually this patient had no problems with the maggot therapy, and was immensely relieved that his foot did not require drastic surgery.

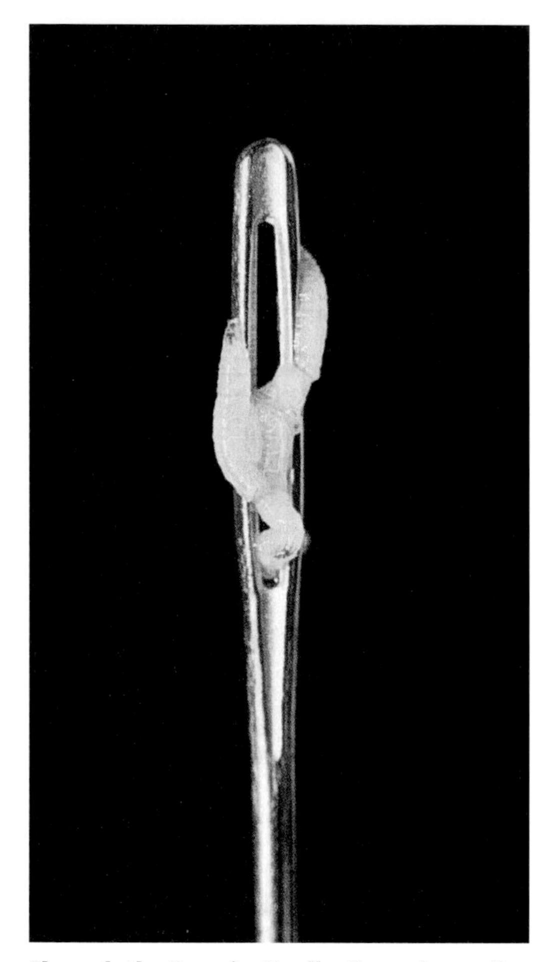

Through the Eye of a Needle. Even when well-fed, the maggots are very small; and when Dr. Steve Thomas applies them to an infected area their activity causes little disturbance.

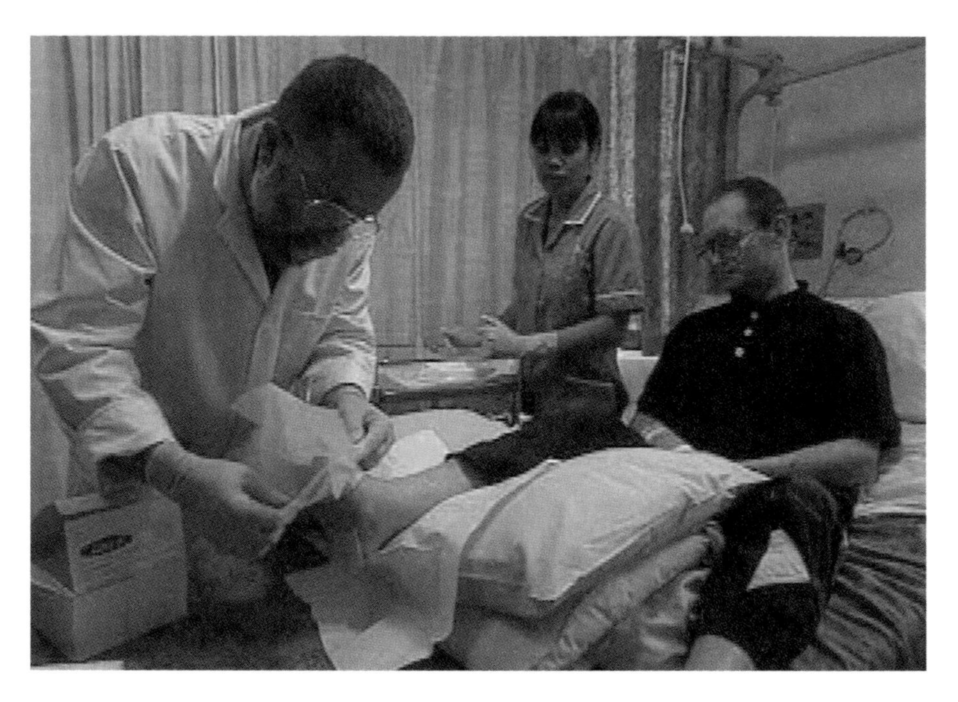

to grow, and the constricted blood vessels prevent the body from delivering antibiotics from the blood stream. On top of all that, some of the bacteria have already acquired resistance to antibiotics anyway, so even if they were delivered to the ulcer, they wouldn't kill the bacteria.

The solution seemed at first to be bizarre and almost crazy. But the fact is that the maggots of the greenbottle fly are actually debris-devouring machines—their entire existence is directed to eating up biological garbage and rot. Steve Thomas spent several years perfecting ways of preparing the live maggots so that they are themselves clean and sterile (and therefore won't introduce more dirty bacteria when they are put on the ulcer). Bred from greenbottles fed on fresh liver (which gives the breeding room a certain aroma, I can tell you), they are then cleaned and sterilized and put in jars ready for use.

When the physician in charge prescribes them, a tiny spatula-full of maggots (containing two or three hundred of them) is gently dropped onto the ulcer. To a distant observer, it looks like any normal procedure of wound dressing in which some powder—an antiseptic or a drying agent perhaps—is dropped onto the area. Except that this powder moves. When you look closely you can see that the powder actually consists of tiny living maggots, measuring 2 to 3 mm in length, that, as obediently as a pool full of seals at feeding time, eagerly cluster round the ulcer and move into it.

That sight was surprising enough, but what the maggots did next was astonishing. As they got to the edge of the deepest part of the ulcer, they all dived nose-first into it, leaving only their bottoms sticking up in the air. Within a few seconds it looked like a neat box of tiny pencils with blunt ends up and (one assumed) the business ends all pointing down. As Steve Thomas explained it, the maggots actually breathe through their bottom ends, fortunately for us, so that they can carry on eating debris and muck without pausing for breath. Once at work, they produce a whole slew of enzymes that breakdown and liquefy the mesh of pus and clot that fills the ulcer bed. As they eat the bio-garbage, they get bigger—so that by the end of their tour of duty they are as long as 8mm (about a third of an inch).

The million-dollar question is a simple one: How do the maggots know when to stop eating after they've cleared up the debris, and why don't they carry on eating up the patient's whole foot? Steve Thomas enthusiastically explained that there just happens to be a defense system built into human tissues that stops the maggots scavenging abilities stone cold. The enzymes that the maggots use to digest biological muck are themselves neutralized by enzymes produced by healthy human tissues. Hence, when the maggots eat their way down to healthy tissue, they simply run into the equivalent of a brick wall. They can go no farther. Not all animals have tissues that produce those neutralizing enzymes, so if we were rabbits, for example, the maggot might indeed eat our hind-leg off.

Macabre but Rather Reassuring Thought

The fact that maggots eat only dead material works to your advantage—provided that you are alive.

I spoke to a few patients about their experiences with the maggot treatment—and they were wonderfully enthusiastic. One patient explained that, in the first place, he was prepared to tolerate almost anything rather than have an amputation and, in the second, when he felt the tiny amount of tickling that the maggots produce, he imagined it as a tiny Fifth Cavalry thundering very quietly to the rescue. None of the patients had feelings of revulsion or aversion. Perhaps more than any centre I visited, Steve Thomas' unit at Bridgend gave me the feeling that knowledge and the intelligent application of that knowledge can really transform something that initially provokes a real "Yecch!" response into the very model of the world's smallest surgeon.

A Worms' Eye View of Our World

But perhaps the most exciting aspect of the world of worms—and I do realize that the word "exciting" is a peculiar one in this context—is the

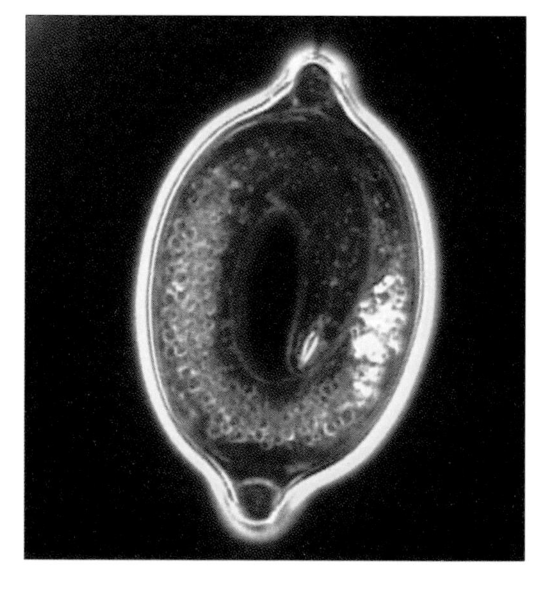

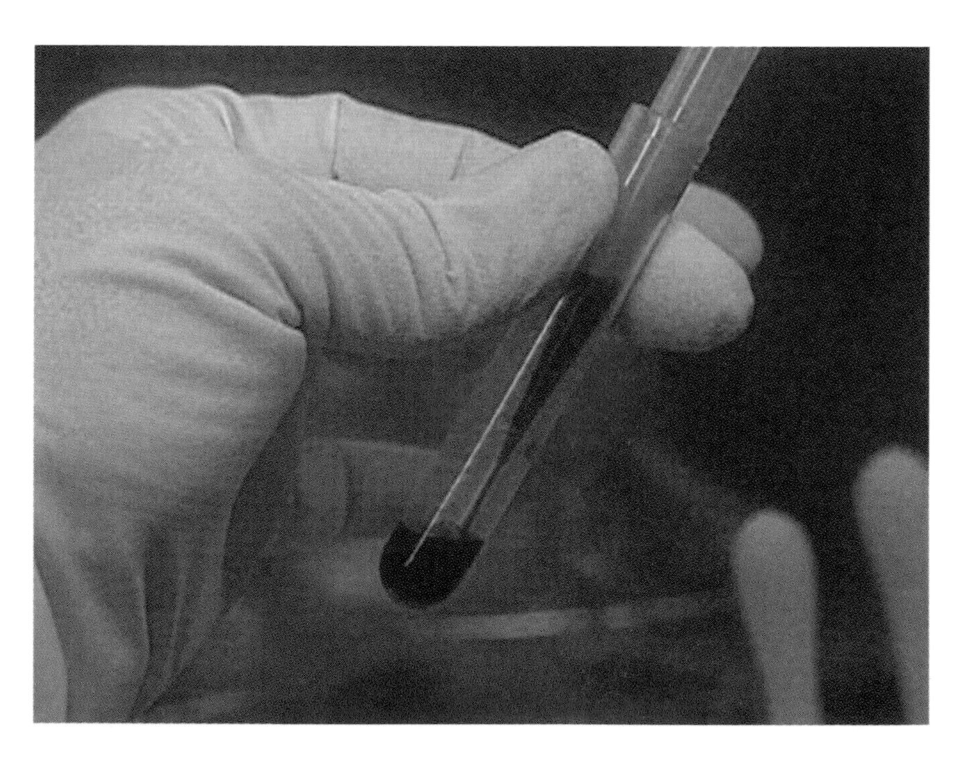

Take Two Thousand Worm Eggs and Call Me in the Morning. A dose of *Trichuris* is prepared for use.

relationship between worms and the development of the human immune system. As we saw in Chapter 3 with the oral flora (and as we shall also see in Chapter 6 with the colonic bacteria), the human immune system may actually benefit or even require exposure to other species at an early stage of development in order to mature and function effectively.

The role of worms in this process is still poorly understand and much of the research is still quite young, but early findings suggest that there may be something very important here. At the University of Iowa, in Iowa City, Iowa, I met Dr. Joel Weinstock and Dr. David Elliot, whose area of research involves the basic interactions of worms and the immune system and the exciting possibility that there may be important implications for new forms of treatment. The idea perhaps began with the findings of Dr. Eric Ottesen, who looked at various communities on the distant island of Mauk. He noted that as public health improved, the incidence of various types of allergies went up. Specifically, he noted that when

(Left) Dr. Joel Weinstock
(Right) Dr. David Elliot

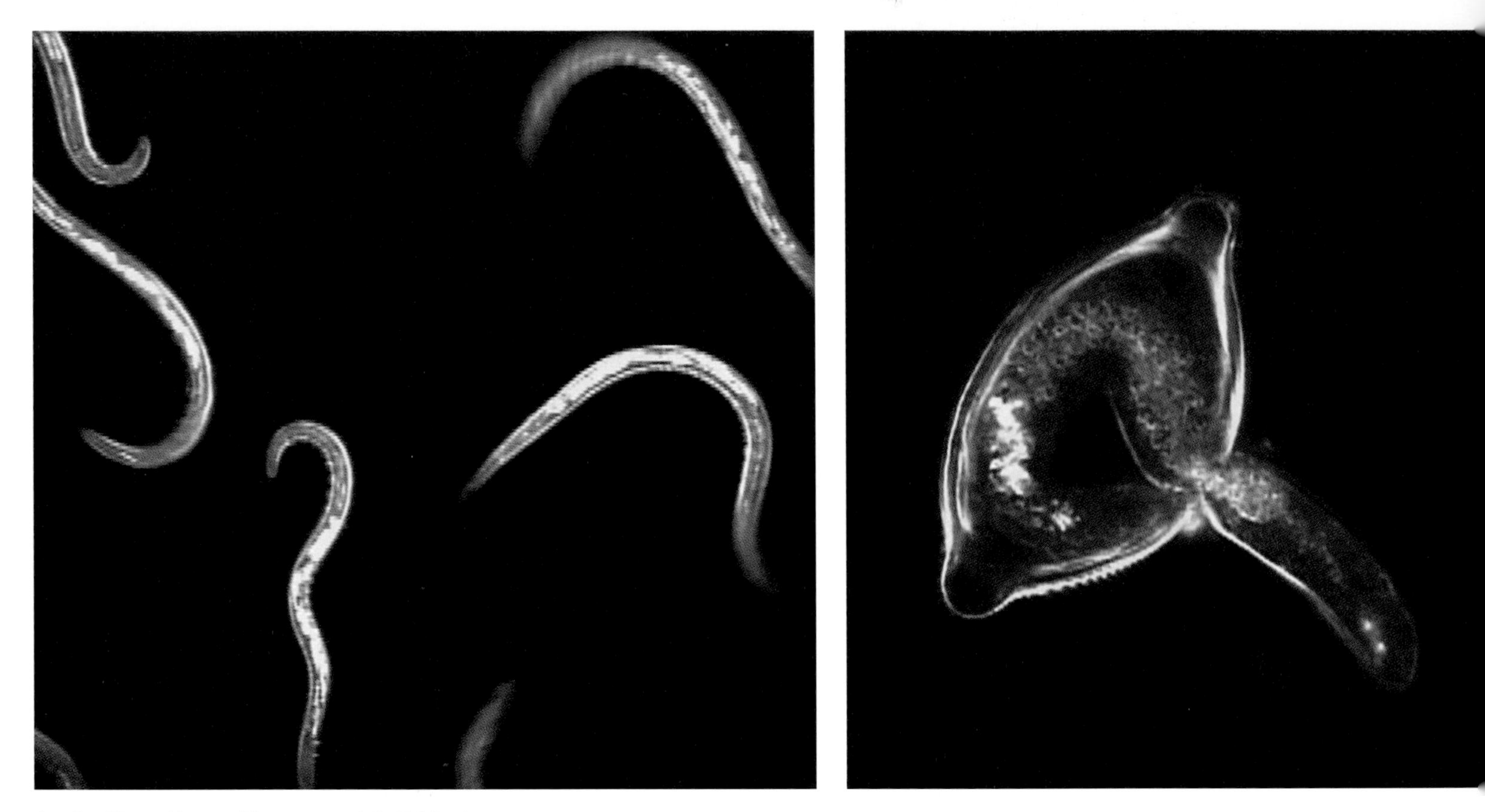

As the Worm Turns. The young adult *Trichuris* worm cannot establish itself inside a human—but it may be able to help with immune-mediated conditions even be making a brief visit.

communities lived in much cleaner conditions and did not suffer infestations of worms in early life, they later suffered more often from allergies such as asthma and some bowel problems.

Joel Weinstock and David Elliot have now done some major research in this area—and Weinstock, together with some colleagues, has started some exciting clinical studies. The theory is fascinating. When we mount an immune response to an invader—an infection, for example—our body's immune system manufacturers several different types of antibody that bind to the same invader. It is almost like attacking an invading tank with a hand-grenade that homes in on it, following up with a rocket-grenade that goes for the same target and then following that with a "smart bomb." In the case of the immune system, there are three major classes of antibody called IgG, IgA, and IgM. However, when the invader is a worm, the body tends to manufacture a class of antibody we call IgE. (Just for the sake of completeness, one other type is IgD, but we don't need to bother about it now.)

Drs. Weinstock and Elliot have evidence that there is something special about a person's early exposure to the type of invader that triggers an IgE response. They believe that somehow the early IgE production tunes up the immune system and makes it more stable. It is almost, according to Joel Weinstock, as if there were several "off switches" in the immune system that had to be properly turned off to stop the immune system from being jittery and unstable. One way of making sure that those switches are in the "off" position, according to this theory, is exposure to certain kinds of worms and the induction of an IgE response. The proof of the pudding—as the saying goes—is in the eating—and in the satisfactory digestion of that meal.

One condition that seems to be based on allergic response is Crohn's disease—a condition in which the patient has intermittent episodes of colicky abdominal pain, diarrhea (often with blood), and changes in bowel habits. Tests of the bowel, often including biopsy, show a characteristic pattern under the microscope. It is not a common condition and can be very debilitating.

It has long been thought that allergic responses are the underlying cause of Crohn's and so Weinstock and his colleagues began a carefully structured clinical investigation. They tried treating the disease with the eggs of a worm called *Trichuris,* a roundworm often found in pigs. The beauty of using *Trichuris* is that—because it is not normally a parasite in humans—when a person swallows, say, 2,000 *Trichuris* eggs, none of them can establish themselves, lodging inside the bowel and maturing into adult worms. The eggs will hatch, but cannot do anything more than that and a few weeks later there is nothing left of them. (If the patient were a pig, it would be a different matter.) However, by then the work is done: the patient's immune system gets a challenge and responds with a flurry of IgE manufacture. Early results are very encouraging indeed. After taking regular doses of the *Trichuris* eggs every few weeks, many of the patients with Crohn's disease have noted marked improvement in their symptoms and weight gain.

Weinstock and Elliot are fascinated—and fascinating. They point to evidence that there may well be several other diseases that can benefit from a worm-induced "damping down" of an over-excitable immune system—studies are already being planned in certain kinds of arthritis and in multiple sclerosis.

We live in a pretty clean and safe world—and, while that's very good in some ways, it may have a down side. Our houses are almost sterile, we eat clean food, and drink clean water—and as a result we don't die of cholera or typhoid or plague. But our immune systems may have evolved to accommodate—or even rely on—early exposure to a variety of wildlife, including worms. The studies of Weinstock and Elliot may have important and far-reaching implications for the future—perhaps twenty years from now little children will be given a few doses of *Trichuris* or some other harmless worm to kick-start their immune systems into a stable and steady state. If so, future generations will certainly be tipping their hats to Joel Weinstock and David Elliot.

Repulsive Invaders

Complex symbiotic relationships are a signal feature of every rich and delicate ecosystem, particularly tropical rain forests and human armpits.

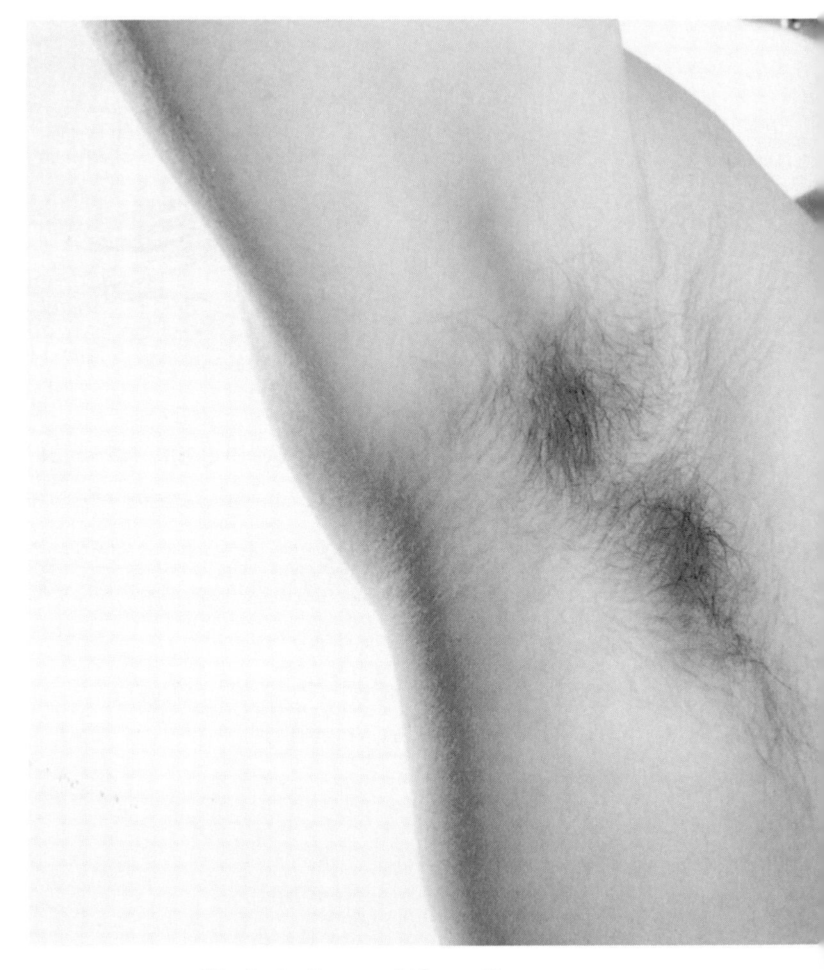

The Rain Forest of Planet Human.

Dank, Dark, Humid, and Teeming with Life

Most of us have a fairly strong image of what a rain forest would be like—even though we may not have visited one ourselves. We imagine—correctly—that it would be extraordinarily dark, dank, and fertile. We can almost hear the gentle dripping of the rain—tropical rain—as it percolates down through the lush green canopy to the forest floor, which is so warm and humid that everything is encouraged to grow. We fondly imagine that if you left a walking stick in the soil, it would be in bloom by the end of spring.

You would imagine all that, wouldn't you? And, as it turns out, you'd be absolutely right.

To investigate the delicate ecosystems of the humid and dank tropics, I went to Costa Rica (oh, the sacrifices I am prepared to make for the sake of my art and my public!). And I have to tell you that a tropical rain forest is just as wonderful and magical and mysterious and quietly fertile and fecund and enveloping as you could hope. However, I had spent so much time reading up on the various ecosystems in various areas of the human body that when we were no farther than a hundred yards into the forest, my background reading got the better of me and I found myself thinking only one thought: "If I were a bacterium living on a human being, I bet this is what an armpit would look like to me (apart from the colour, since most armpits, although lush, are not particularly green)."

It's probably true—to your average *Corynebacterium* living around the hair roots in the average man's armpit, the environs would surely seem just like a rain forest. But what I did not realize until I began the research into this topic was how important the denizens of the axilla are in affecting human behaviour. To put it simply, the bacteria in our armpits are not merely visitors or casual

Reassuring but Totally Impractical Fact

Fresh sweat has no smell.

. . . and the Rainforest of Planet Earth.

Whether on a human or the larger jungles of our planet, both areas are moist, warm, and have an aroma all their own. Both are home to many types of wildlife and play an important role in all of our lives.

tourists—they work there. It is the bacteria in the armpit that actually create chemical signals from the raw materials in the sweat that we produce—and these signals influence, as we shall see, some important aspects of the behaviour of the people around us.

Preti's Voyages. Intrepid expeditions by Dr. George Preti bring back live specimens of *Corynebacteria* from the hinterlands of a volunteer's armpit.

"Let Your Armpits Be Your Charmpits"

Dr. George Preti of the Monell Chemical Senses Center in Philadelphia spends his day with a pipette poised in the armpits of volunteers. He studies the sweat they produce (hey, it's a living—AND it's good science) and the signals that are transmitted in that sweat.

Let's start with some basics. Sweat itself doesn't smell. I know that sounds wrong, but actually it's true. Although we all know when something smells "sweaty," that smell is not intrinsic to the sweat itself as it is produced. (By the way, this came as a complete surprise to me—so don't feel discombobulated if you didn't know that either.) The smell is produced by the activity of the bacteria in the armpit, which are predominantly members of the species *Corynebacterium,* as they work on the various molecules in the material they find in the armpit—the sweat itself and *sebum,* the oily secretion of the sebaceous glands at the base of the hairs.

The *Corynebacteria* essentially live off the sweat and the other secretions in the armpit—as the saying goes, it's breakfast, lunch, and dinner to them—and they change it in two ways. First, the enzymes that they manufacture—called *lipases*—break down some of the fat molecules present and produce small molecules that have a distinctive property:

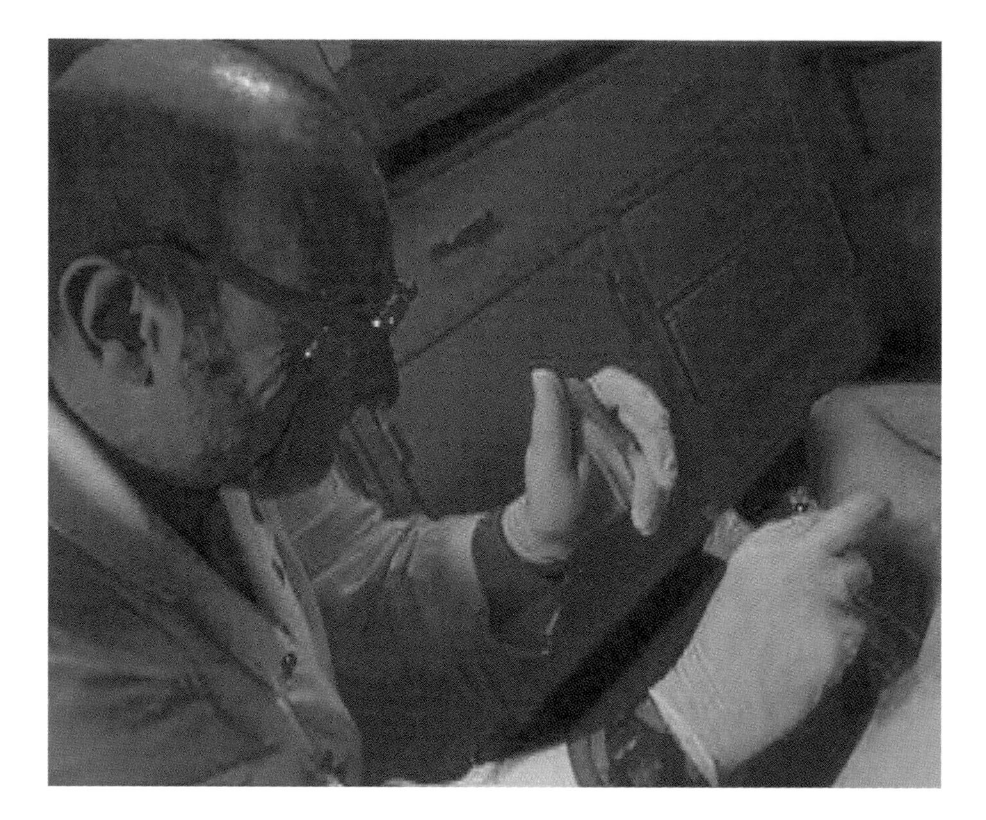

Scenes from a Peculiar Workplace. Dr. George Preti, an international authority on body odour and the chemistry of sweat, is seen here gathering some important material for his research.

they smell rancid. (I believe the correct term in organic chemistry is "aromatic." I once visited an organic chemist's home and a lot of things in his kitchen smelled pretty rancid and pongy to me, but he said they were aromatic, so perhaps it is a matter of professional euphemisms.) Furthermore the exact mixture of "aromatic" compounds—the recipe that makes up a person's metabolized secretions—seems to be individual and characteristic of that person.

A famous experiment was done at the University of Berne in Switzerland in which a group of men volunteers had to wear T-shirts for two days during which they were not allowed to take baths or use deodorants. At the end of that time, the T-shirts were placed into unmarked "sniffing boxes" and the men's girlfriends had to sniff each shirt in turn and try to identify their own boyfriend's shirt. Which they did. The success rate was very high—proving that the smell of a person's

Who Needs a Bloodhound? In a re-creation of the famous University of Berne experiment, girlfriends were able to identify their boyfriends by the smell of their T-shirts.

armpit (and probably chest) is fairly individual and characteristic of that person. (In fact there may be even more to it than that—it may be more than the smell, as we shall see in a moment.)

So that's the first thing that *Corynebacteria* do—the bacterial wildlife creates a smell from the odourless sweat. The second thing they do is even more surprising—the bacteria also create the molecules that send chemical messages out to other people—the molecules that we call *pheromones* (meaning "hormones that work outside the body"). Now, the history of pheromones is quite interesting—and I shall give you the briefest of overviews. Basically, it all started from some famous observations by Martha K. McClintock at the University of Chicago in 1969, who noticed that, after several weeks of living together in a dormitory, women synchronize their menstrual cycles, and—having started their cycles at different times—end up all menstruating at the same time. There were also a number of anecdotal reports that many young women who had attended certain rock concerts at the same time had the same experience. These observations suggested that there was a chemical signal coming from some of the women that influenced the menstrual cycle of the other women. Phenomena like this had been known in many animals—chemical signals are, for example, used by ants to "mark" a trail for the worker-ants. In the case of the synchronizing female roommates, further studies have shown that these chemicals originated in the armpit secretions of the women—in other words, they were made there.

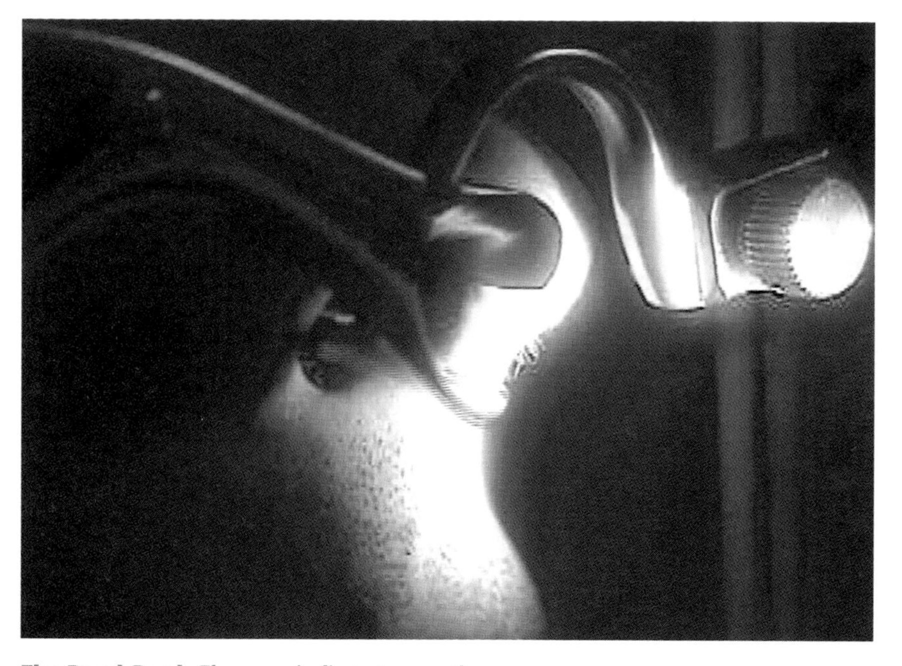

The Royal Road. The nose is first stop on the road for pheromones, which have many effects on different systems in the recipient's body.

Dr. Preti's group then showed that female menstrual cycles could also be influenced by secretions from male armpits—the neighbouring males could influence nearby females by their pheromones. Furthermore—and this seemed to me to be so unusual that I had to ask Dr. Preti four times before I fully believed it—the pheromones are not secreted in ready-to-use form by the glands in the armpit, they are produced by the action of bacteria on the person's sweat. In other words, human pheromones are not products of human beings, they are products of a human-plus-wildlife conglomerate. If we had no bacteria in our armpits, our sweat would just be sweat and we would not produce pheromones. (This, by the way, has been shown by careful and detailed experiments and studies on the chemical components of sweat by Dr. Preti and his group.)

Now, while we are on the subject of pheromones, we can consider some of the newer and less well understood areas of current research. When the discussion of pheromones first started, a lot of people thought that pheromones would turn out to be major controllers of human behaviour. They envisioned worlds where pheromones would be sprayed over a crowd to make them docile and peace-loving, or would be sprayed over soldiers to make them aggressive and lethal. So far, there's no evidence that this really happens in humans. It does in some animals—there are pheromones that alter the behaviour of rats that can produce both mating behaviour and aggressive behaviour that can only be reversed by a quick whiff of the right pheromone. But, to date, that kind of control function has yet to be demonstrated in humans. It is possible that this may happen, but it hasn't yet.

Another interesting bit of research shows that pheromones may also contain genetic fingerprints (in the form of certain molecules that are unique to the individual). There is early evidence that this may produce changes in mating behaviour (for example, females may deliberately seek out males whose DNA is different from their own father's—thus expanding the gene pool). Of course, these are early days in this type of research—but watch out for more findings from Preti's group and co-workers.

The Bliss of the Kiss

While we now realize that the bacterial wildlife in the armpit can change the nature of the smell and appeal of the individual, the same is also true of the smell of the breath. Although it seems that the breath does not contain any pheromones that actually influence the behaviour of the people around us, the odour of the breath—and the taste when it comes to kissing—can make a great deal of difference.

Dr. Alan Hirsch has investigated this extensively in Chicago and his conclusions are somewhat unexpected. According to Dr. Hirsch, the nose, in many respects, is a more important sexual organ than the penis (this is debatable in some ways, but you can

Leading by a Nose. . . Dr. Alan Hirsch supports his contention that the nose is more important as an organ of sexual arousal than the penis by comparing models of each. (*Warning:* Objects may be smaller than they appear in the mirror. Or anywhere else for that matter.)

see what he means). He has measured changes in blood flow in the brain when a subject sniffs in various smells and you can see on the screen of his scanner what happens as the subject's brain registers an interesting smell and the information is analyzed and considered. The analysis of the smell begins in the frontal lobes—associated with emotions in general—moves on to an area called the *nucleus acumens* (associated with pleasure), and then to the *septate nucleus,* which is associated with sexual arousal. In addition to these more remote-control centres, Dr. Hirsch also did studies in which he measured changes in the blood flow to the penis. Should you ever be asked about this, the correct term for that test is *penis plethysmography*. Now you know.

Hirsch's results give pause for thought. When he tested over a hundred young males in Chicago and measured their brain responses and changes in their penis plethysmography, he found that sexual arousal was stimulated by... well, by every single odour that he tested. There was nothing that wouldn't arouse a male adolescent (which, in my opinion, explains a lot about the way the world works). However, of all the odours he tested, the ones that gave the greatest arousal to the males were the smells of lavender and of pumpkin pie. Now that's a difficult result to understand. I mean we all like Thanksgiving to some extent, but, honestly, by the time we've cleared the turkey bones and set out the pumpkin pie, are we really ready for serious sexual arousal? Perhaps things are different when you have Thanksgiving dinner at the Hirsch's place. Particularly if you've got a plethysmograph in your underwear.

Information from Way Over in Left Field

While males are sexually aroused by virtually any odour, the ones that work best are lavender and pumpkin pie.

The female of our species seem to exercise much greater selectivity—they were only aroused by the smell of licorice, cucumber, and banana nut bread (I am NOT making this, up by the way). Furthermore, female arousal was inhibited by the smell of barbecued meat, cherries, and male cologne. So if you thought you were going to do well by being a sensitive guy and barbecuing the dinner while wearing cologne, think again. As Dr. Hirsch put it, you'd be better off throwing away the cologne and getting a cucumber. Unexpected advice, perhaps, but doubtless significant.

So much for smells. Hirsch also did a large series of studies on the effects of various tastes on human sexual arousal. Here the experiments were perhaps a little less sophisticated. Dr. Hirsch placed a drop of a

Bizarre Information that Sort of Makes Sense

Female sexual arousal is inhibited by the scent of barbecued meat and men's cologne, but is stimulated by the smell of cucumber.

Facts that Certainly Explain Almost Everything

1. Married women prefer their husband's kisses to taste fresh like toothpaste.
2. Single men like their date's kisses to taste of alcohol.

flavoured liquid on one person's tongue and then asked him or her to kiss another volunteer, who then recorded her or his impression of the taste and attractiveness of the kiss. Among females, the strongest preference was for things that tasted like toothpaste—peppermint and spearmint. Furthermore, both married women and single women had the same preference, so the breath mints really do work.

Among men, however, there was a significant difference between the married men and the single men. The married men preferred the minty fresh tastes, but the single men liked the breath of their kissee to be flavoured with alcohol. I'm not sure what conclusion we can draw from this research, but it does suggest that alcohol plays some role in the dating process—though not necessarily in the way we thought.

Acne—Wildlife in a Tight Spot

The really embarrassing thing about acne is that it is a really embarrassing and highly visible skin condition that most often occurs at a stage in life when we are really embarrassed about things that are really embarrassing. Such as acne.

The second significant thing about acne is that almost everybody is absolutely certain that they know what causes it, what triggers an outbreak, what to do to fix it, and what everybody who has got it is doing wrong. Myths about acne are commoner than myths about any other human condition. And they are just that—myths.

Most people imagine that acne is caused by eating chocolate, eating too much fatty or fried food, too frequent meals of spicy foods, not washing the skin properly, staying up all night listening to music, or getting nervous about a romantic date. One of the teenagers I spoke to while preparing this section said, "Being a teenager is all about the four P's—pubs, pizza, puberty, and pimples." Well, not quite. To generate acne you definitely don't need pubs or pizza, nor yet popcorn, pimentos, paprika, pepper, pop music, or pot. The chief culprit is one particular P—*Propionobacteria,* helped along by another P—puberty. The whole thing starts in the glands at the base of hair follicles. These are called

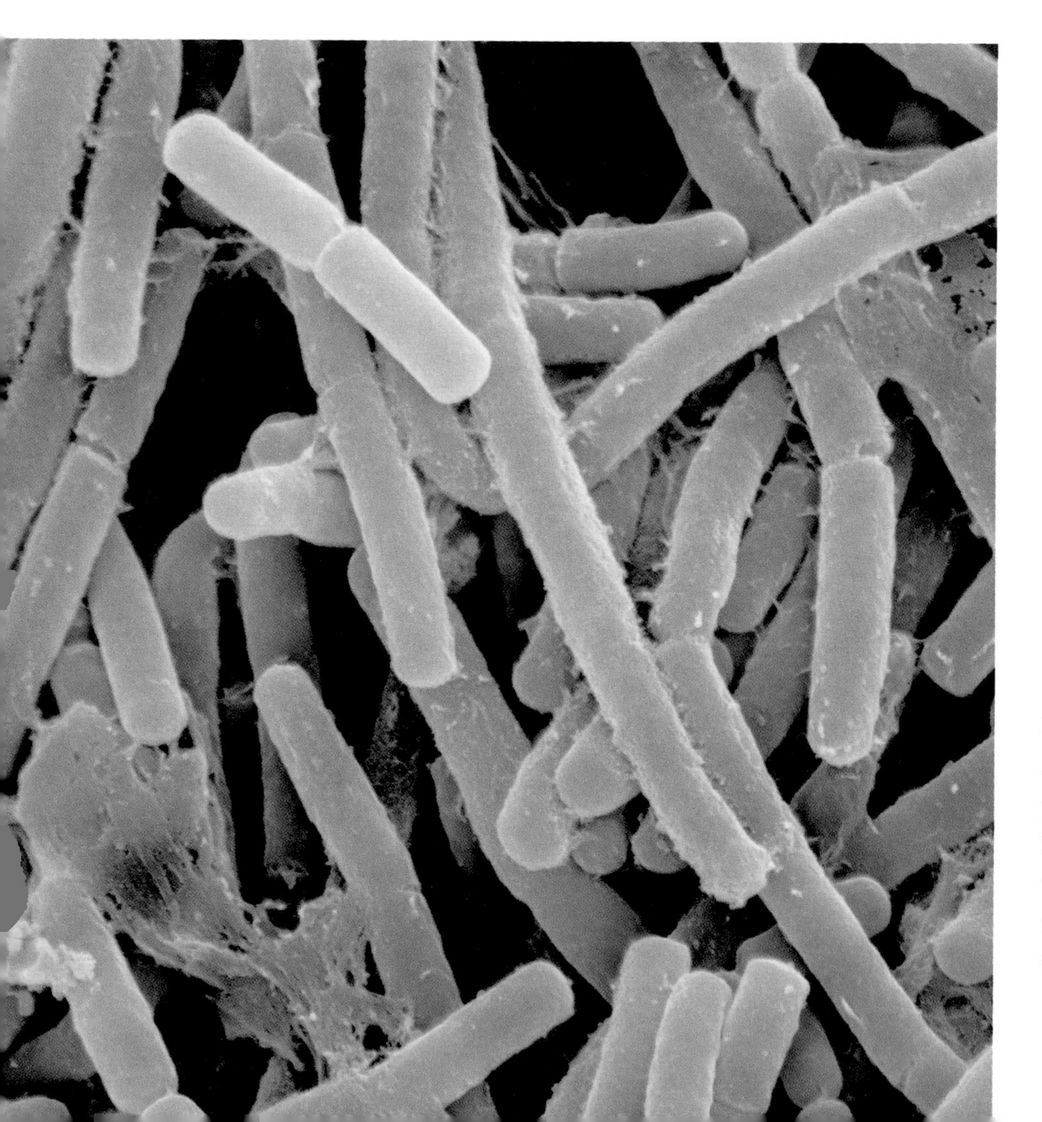

Not So Mellow Yellow. The sebaceous glands are situated at the base of hairs. The sebum they produce helps to keep the hairs waterproof. Acne begins if the duct gets blocked and these characters then arrive. Looing like a close up of macaroni and cheese, these *Propionobacteria* feed on the sebum and help create the contents of a zit.

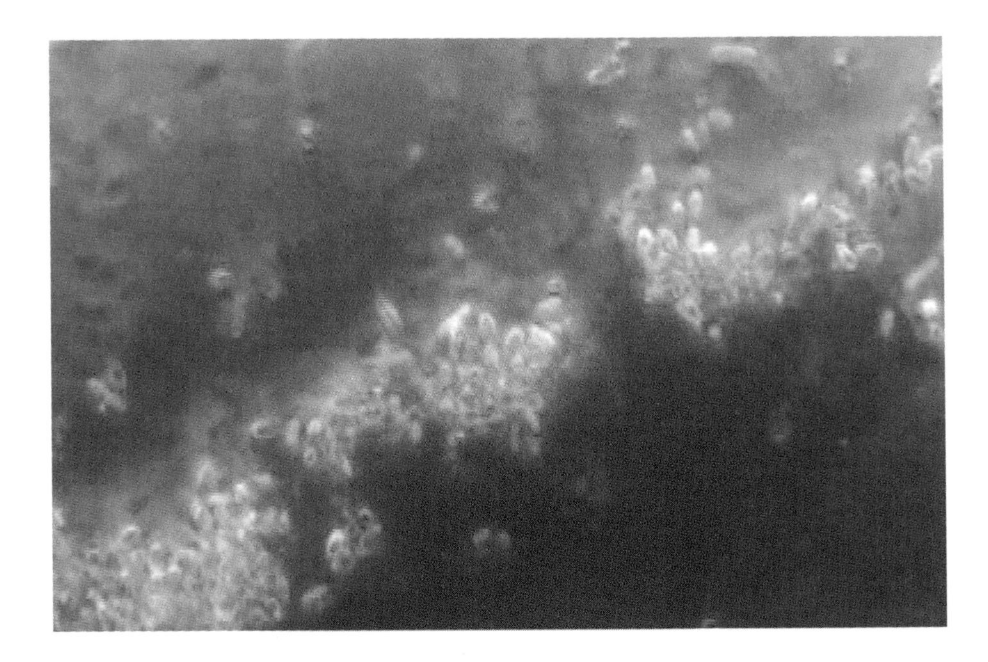

***Propionobacteria* at Work.** Tightly clumped together and feeding on sebum.

sebaceous glands and their function is to secrete an oily substance called sebum, which is—among all mammal species—a waterproofing agent that allows skin covered with hair (or fur, as they know it) to keep relatively dry when the owner is swimming or is caught in the rain. The sebaceous glands cluster around the root of the hair shaft and manufacture the sebum, which gets onto the hair via a duct leading out of the gland.

And it is the duct that is the problem—because if you happen to have narrow ducts, when your glands start making lots of oil (as they do under the influence of testosterone at puberty), the duct just can't handle the volume and gets plugged. The plugged duct is the start of acne. The oil fills out the glands, and the *Propionobacteria* that already live there get to work and breakdown the oil. The breakdown process first changes the colour of oil from clear to black (it isn't dirt) and then makes the whole environment more acid, which starts an inflammation. So what starts as a plugged duct (a whitehead) soon becomes a duct with altered oil (a blackhead) and then a gland surrounded by inflammation and, finally, secondary infection (a pimple or zit). In severe cases, this can turn into

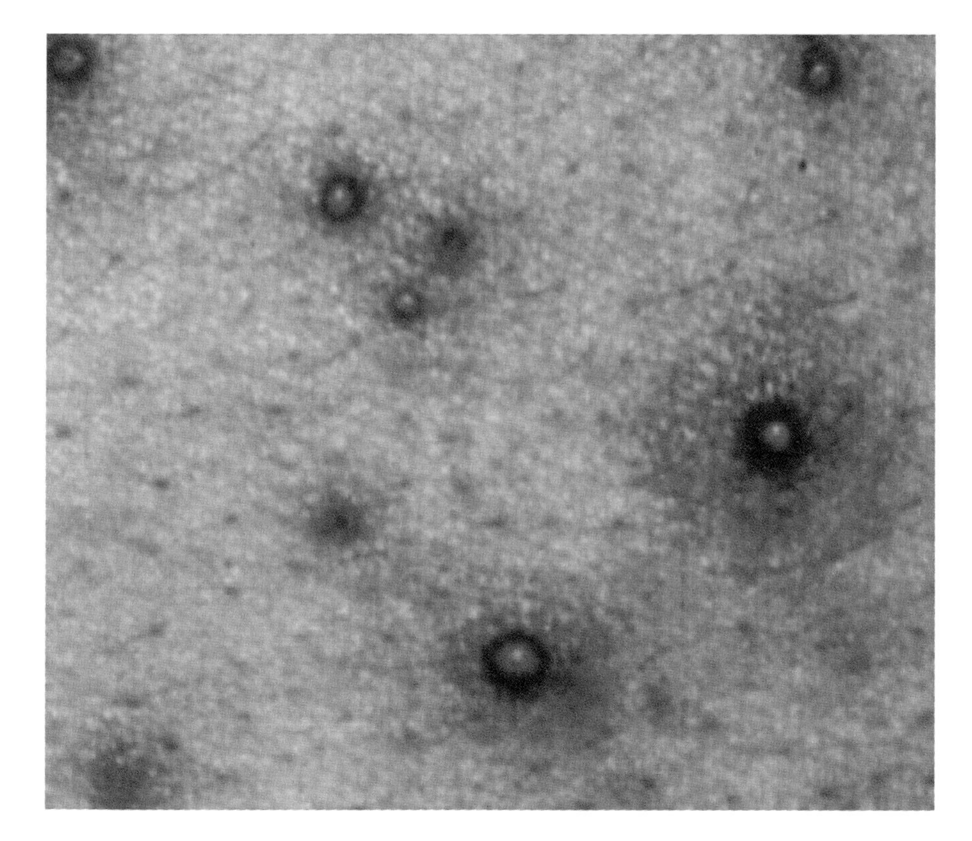

a big cyst (a monster zit) that eventually heals, leaving a sunken scar (a pockmark).

Acne is a classic illustration of our wildlife producing changes that, in turn, alter the way others perceive us. This chapter is called "Repulsive Invaders" and, of course, although most cases of acne are upsetting and do not cause us to be repelled—severe cases do have serious effects. Some studies have shown that really severe acne affects a person's chances of getting a job, and drastically alters a person's social life and mood. So if you have severe acne—or if it is getting that way—please get a dermatological opinion. There are treatments that affect the balance of bacterial wildlife in your skin, and that change the course of acne and prevent scarring. *Propionobacteria* are one example of human wildlife that should be kept in their place if they are causing trouble.

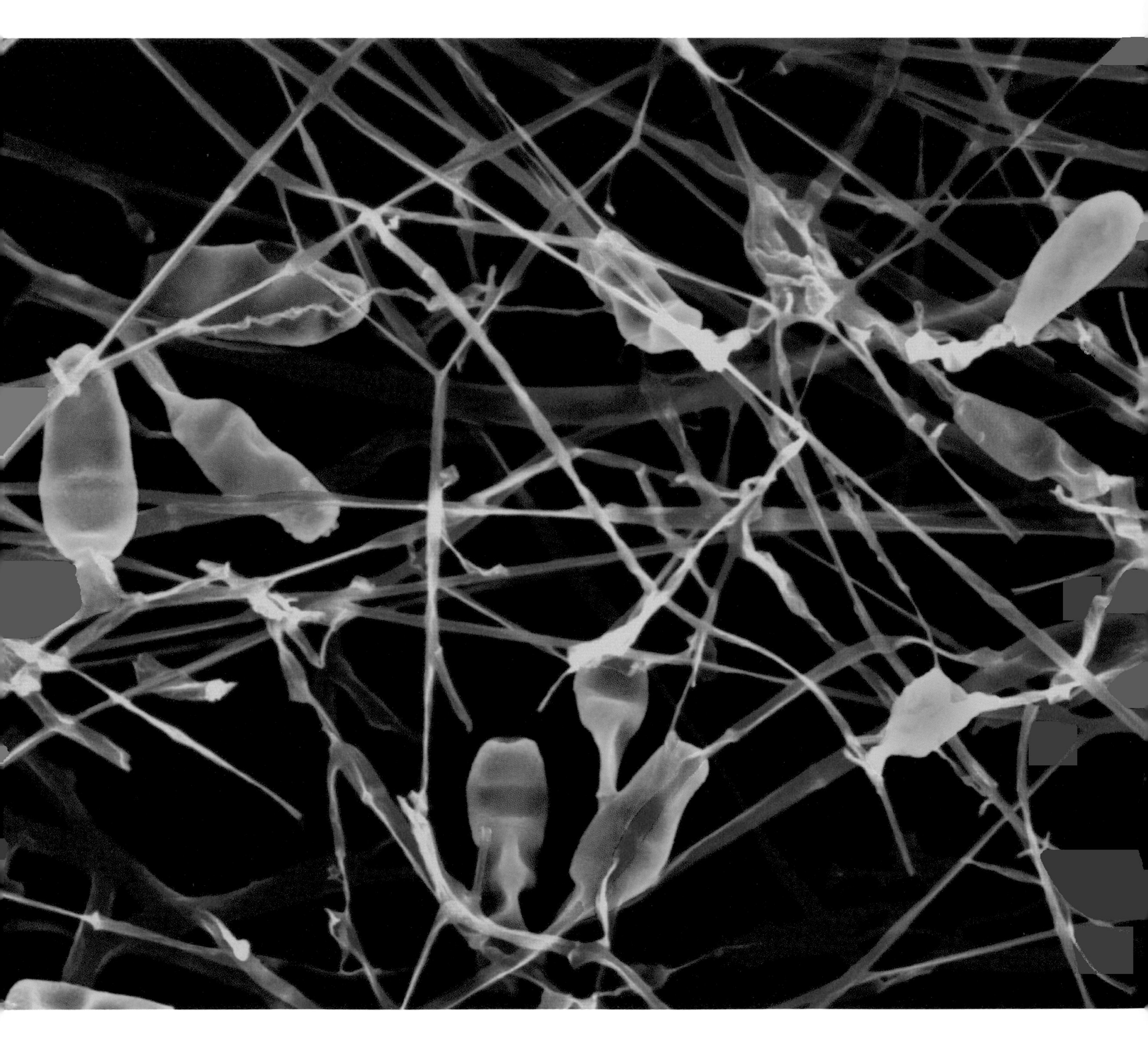

Fungus and Fun Guys

Appearances are deceptive. When you see the delicate structure in the photograph, you may at first think it is a special effect from *The X-Files*. Actually it's much more mundane and pedestrian than that—and pedestrian is the operative word. It's actually a sample of human wildlife from a human foot—it's the fungus that lives in the crevices between your toes. It's known as athlete's foot and, of course, you don't have to be an athlete to get it. All you need is a modicum of sweat and the right conditions in the area —confinement and staleness—to encourage the growth of the fungus, whose proper name is *Tricophyton*.

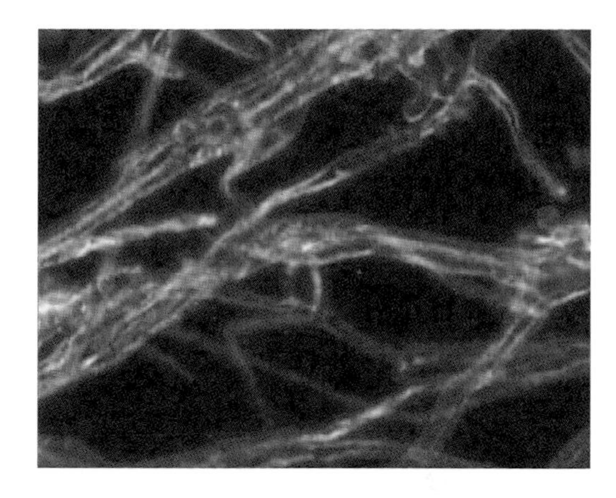

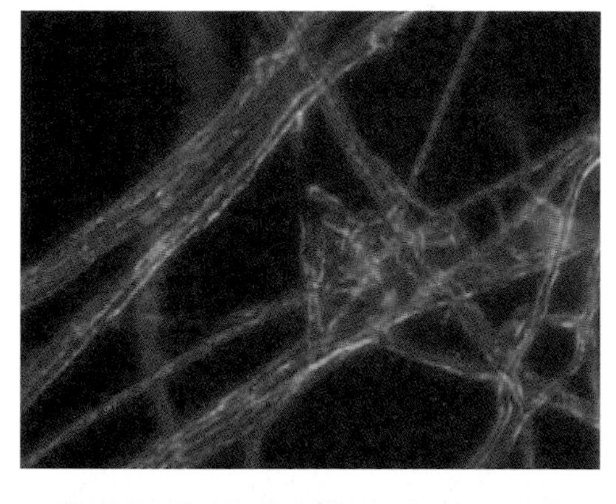

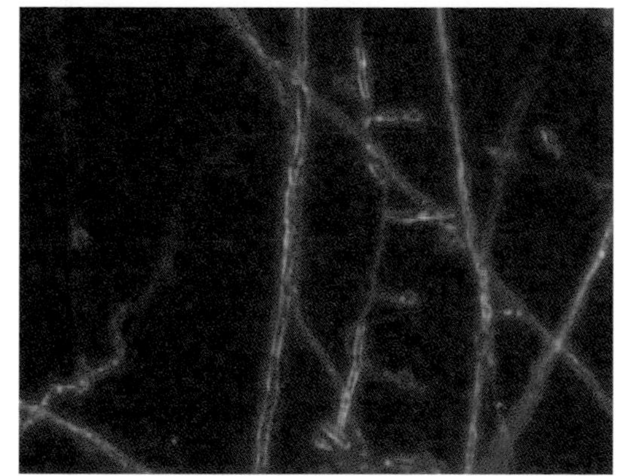

Real Networking. Fungi exist as a delicate interlacing network of branches called hyphae. These don't *connect* the organism with anything—they *are* the organism.

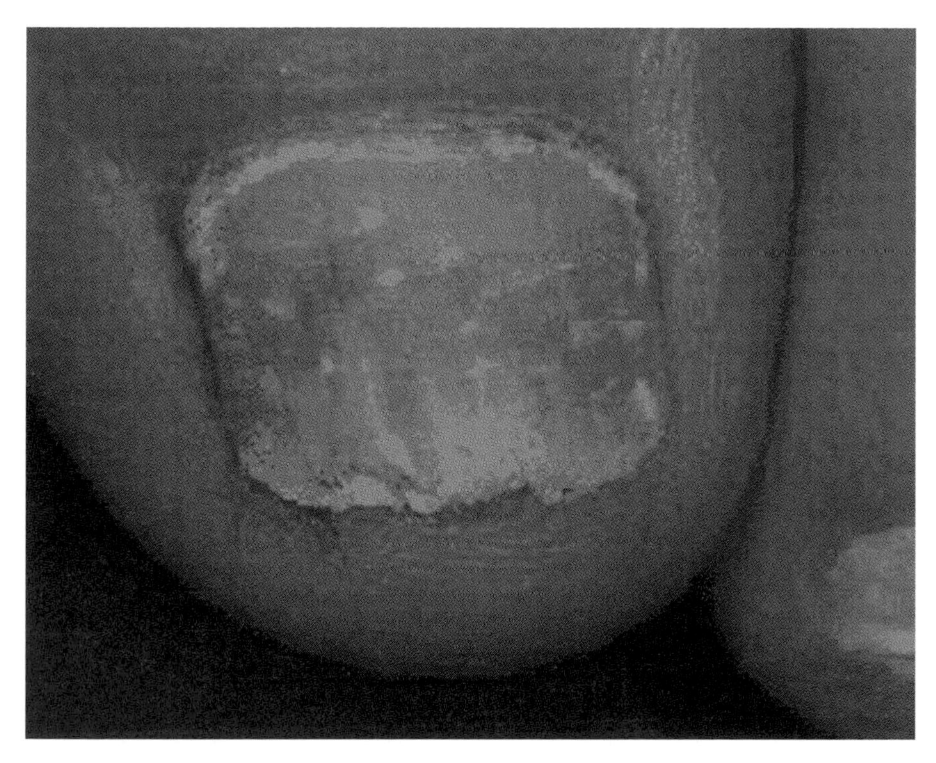

Athlets's Toenail. Sometimes the fungus that causes athlete's foot.

(Opposite)
Epidermophyton Floccosum

Warts and All

According to Dr. Neil Shear, head of Dermatology at the Sunnybrook and Women's Health Sciences Centre in Toronto, if the virus that causes warts had a personality it would be the archetypal guest who outstays his welcome.

A wart is a benign tumour of the skin—a heaping up of over-enthusiastic skin cells—benign in the sense that it doesn't invade surrounding areas or spread to other areas of the body. This tumour is caused by a virus whose full scientific name is *Human Papilloma Virus,* or *HPV.*

This might be a good place to remind us all of what a virus actually is. If you're feeling a bit vague about them, have a look at the information box—and if not, please carry straight on.

To carry a little further Neil Shear's analogy of the wart virus as a mildly unpleasant, loutish visitor who arrives at your house uninvited, settles on your couch, and just won't leave, if you do try to throw him out, he is quite likely to wander back in later and this time settle in the bedroom or the bathroom.

The wart virus is really very contagious. Everybody knows that it can be transmitted at swimming-pools, however, most people don't realize

A virus is not truly a living organism. It cannot live—in any sense of that word—by itself. It needs a living cell in order to produce more of its own kind.

A virus is actually a tiny packet of DNA or RNA that can only reproduce by slotting itself into the nucleus of a living cell and fooling, or duping, that cell into churning out replicas of the virus. It is—if you like—a rogue blueprint, and it can only achieve anything if it can get into a suitable factory and hijack it to manufacture its product. You could almost think of a virus as being like somebody selling something on the Internet—they don't need a shop or an office with all the maintenance equipment and heating bills and so on. In the same way, the virus—because it is really nothing more than genetic information—doesn't need to have all the complex metabolic equipment that ordinary cells do (such as mitochondria, protein systems, and so on).

As a result, viruses are very small indeed, as you can see from the photograph at the top of the facing page. (On that scale, the cell on which the viruses are sitting would probably measure about three feet across.)

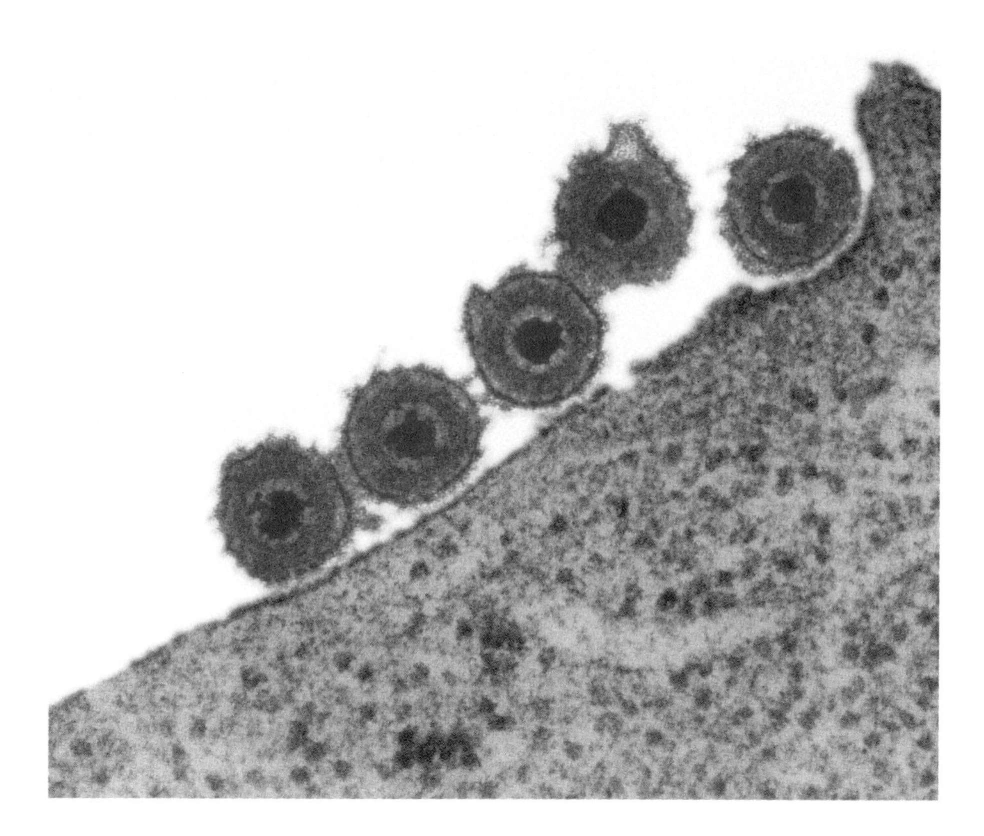

Cold and Sore. This electron microscope picture looks like a few wagon wheels rolling down a hill. Actually they are viruses—a few of the Herpes simplex viruses, the ones that cause cold sores. Once inside the cell—which on this scale would be about three feet in diameter—the virus uses the cell's reproductive system to manufacture more of themselves. It's not fair—but it's the way some wildlife abuses our generosity.

that it is probably not transmitted in the water of the pool, but in the puddles on the side—where people can tread and leave some HPV (if they have a wart) for you to acquire. The virus can also be transmitted from one place to another on the same person. It's quite common for a wart on a finger to be transmitted to the adjacent finger or thumb. There is a

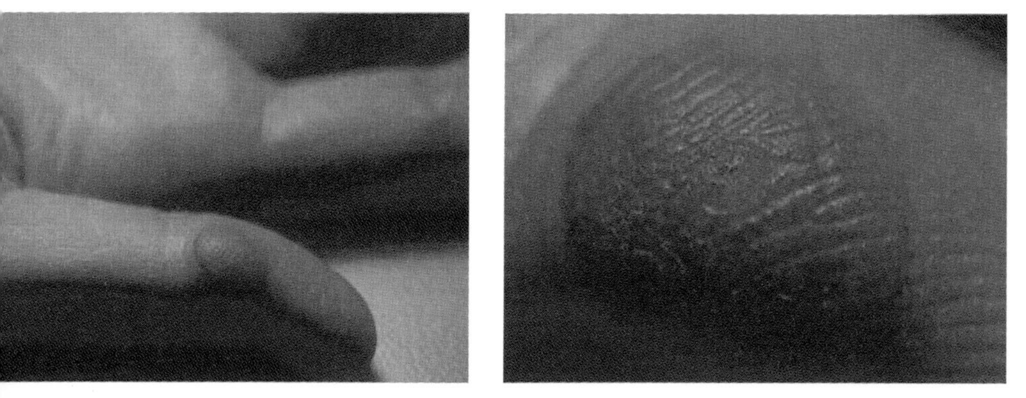

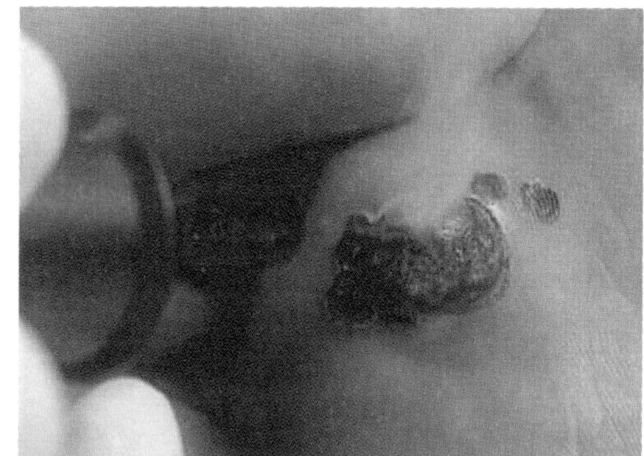

Warts On and Off. Laser treatment is one effective method of dealing with particularly stubborn warts.

very particular relationship between the wart virus and the individual host's immune system. Some people get recurrent warts and some never get them. Warts can be very difficult to get rid of, but can also disappear totally, suddenly, and spontaneously (due, we think, to an immune reaction).

There are literally hundreds of magic incantations, spells, and recipes to get rid of warts that have come down to us since medieval times—including the infallible one of washing in water that has been left under a gallows on which a criminal has been hanged at a cross-roads (a sure-fire success, I'm certain). However, if you happen to live in an area in which they don't often hang criminals at cross-roads, then do consider visiting a dermatologist. It's not as romantic, but it's much more reliable.

What the Cat Should Never Have Dragged In

We are almost alone among the species on this planet in domesticating other animals. The only additional candidate species that I know of is a certain kind of ant that keeps aphids and feeds off the juice that well-fed aphids make—but, although it is definitely a kind of farming, I'm not sure whether that really counts as a type of domestication. Anyway, we alone — as far as I know—domesticate certain animals that do absolutely nothing for us in terms of food supply, but keep us company when we are feeling lonely. The commonest species are, of course, domestic cats and dogs. Which is fine most of the time. Unfortunately, on certain occasions, our pets can give us more than the unconditional love for which they are justly renowned.

Really Good Advice

Take the possibility of *Toxoplasma* seriously—change your cat's litter (and be very hygienic when you do it) every two days, three at the most.

Toxoplasma gondii is the real reason why you should change your cat's litter every two days (three at the absolute maximum). *Toxoplasma* is a protozoan—that means it is a single-celled animal—a parasite that lives in cats without causing much of a disturbance. The cyst stage of the

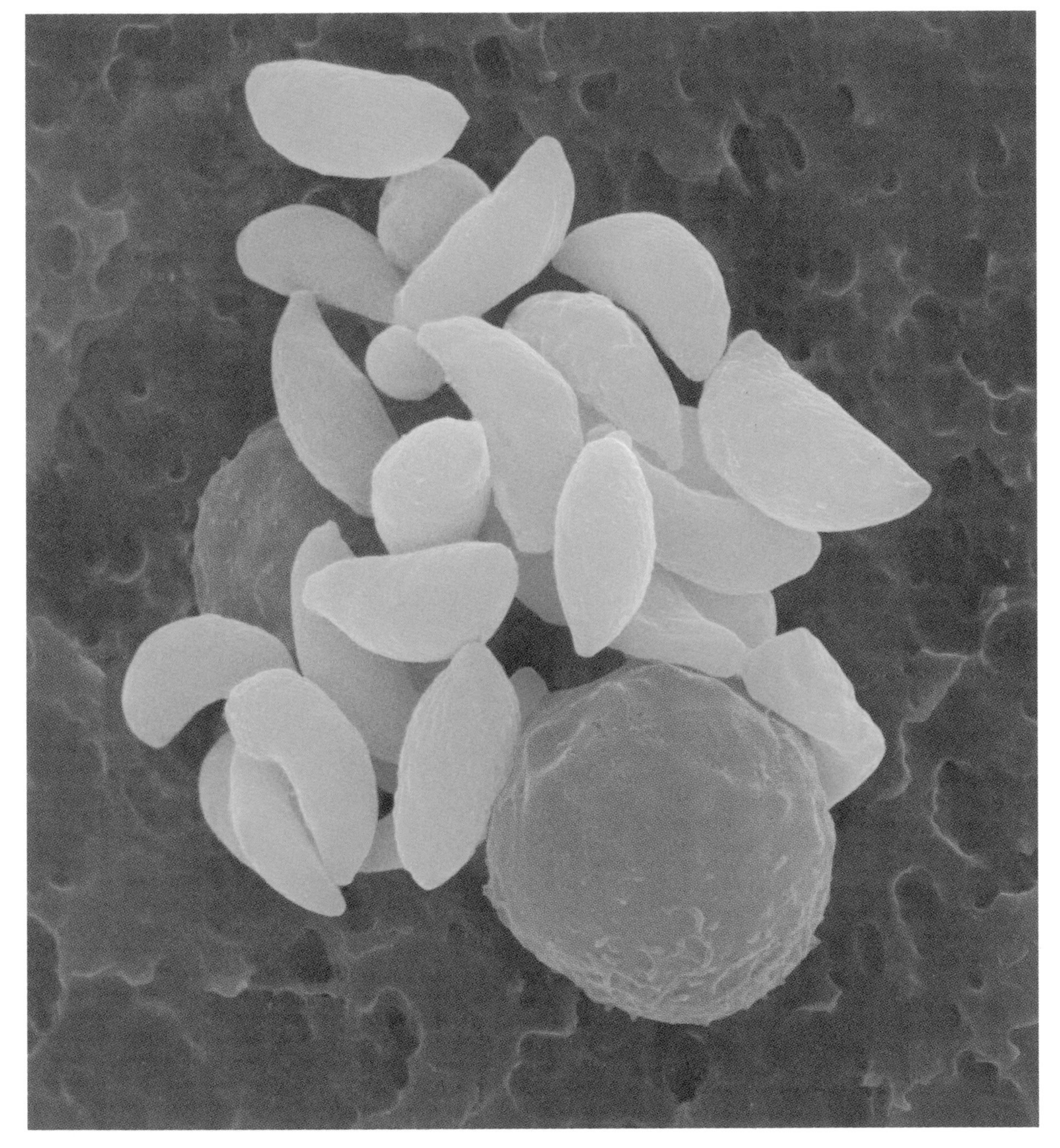

Serious and Subtle Danger. This creature is the protozoan parasite *Toxoplasma gondii*. Ordinarily the symptoms of an infection are very mild, often unnoticeable, but if a pregnant woman becomes infected the fetus can suffer severe and irreversible damage.

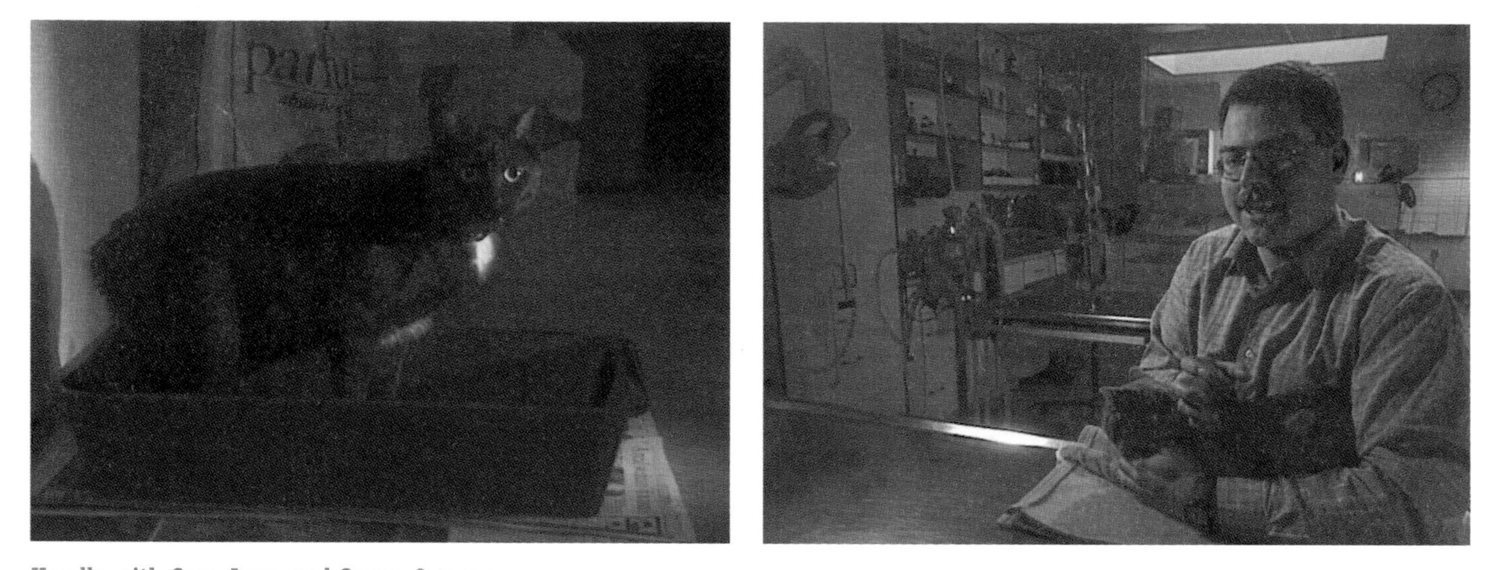

Handle with Care, Love, and Sense. Cats are wonderful and, as Dr. Craig Stephen points out, can continue to be so, provided that we know a few simple rules.

Toxoplasma is passed in the stool of the cat. The protozoan becomes active after about three days and then it is ready to be ingested. Dust from cat feces can be blown about or a child can play near the litter box or an adult can fail to wash carefully after changing the cat litter. If any of these things happen, then the *Toxoplasma* can get into a human host.

In most cases a *Toxoplasma* infection is trivial; in fact, sometimes it can be so mild that the person doesn't even know that they have it. The only danger—and it is a very grave one indeed—occurs if a pregnant woman gets an infection. Then the *Toxoplasma* can do very serious and irreversible damage to the developing fetus, who may be born with many different kinds of defects in the heart, brain, eye, or other areas.

It is the subtlety of the infection in non-pregnant people that makes tracking down the source of an infection so very difficult. There was a particularly tragic outbreak in Victoria, British Columbia. Dr. Craig Stephen, director of the Centre for Coastal Health at the University of British Columbia, and some of mothers who had been infected spoke about what happened. When several children were born with damage, and when blood tests showed that *Toxoplasma* was the culprit, it became quite a challenge to locate the source. As it turned out, in this case, the most likely source was probably not domestic cats, but an area in which a river changes direction—a watershed—where it seems probable that wild (feral) cats may have deposited *Toxoplasma*-infested feces in a park area.

Why We Should Change the Cat Litter Often. The cysts of the Toxoplasma take three days to develop into the next stage—if you throw them out before then, they won't be able to infect you or your family.

The kind of infection that likely occurred in British Columbia, transmitted from wild cats, may be virtually unpreventable. But other precautions are still of great importance. It is extremely important to be careful when changing cat litter (every two days, three at a maximum), handling the litter as little as possible, washing your hands very carefully afterwards, and advising pregnant women to be very careful in visiting houses where there are cats (or even alleyways and sandpits where cats might roam).

The Art of the Fart

At the opposite end of the spectrum of seriousness among humankind's afflictions, there lurks a tiny and totally trivial question: Why is farting such a socially unacceptable activity? It is, after all, very similar to belching, except that the portal of exit is less visible. The main difference, however, between a belch and a breakage of wind is that with the flatus—and this is where the activities of the human wildlife in the colon are significant—there is always the possibility of an unpleasant and pervasive smell. I have always thought that this issue was so important, and so funny (after all I was born in England where it is generally accepted that anything to do with farting is the funniest topic imaginable) that

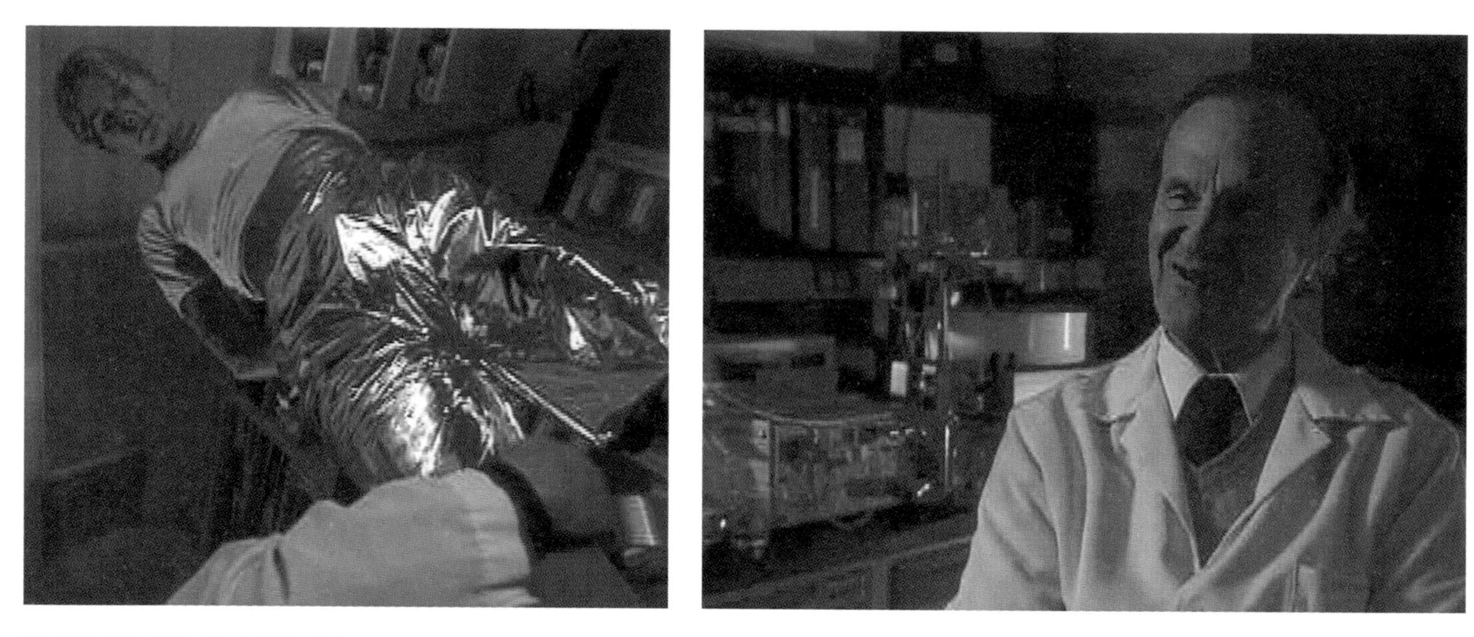

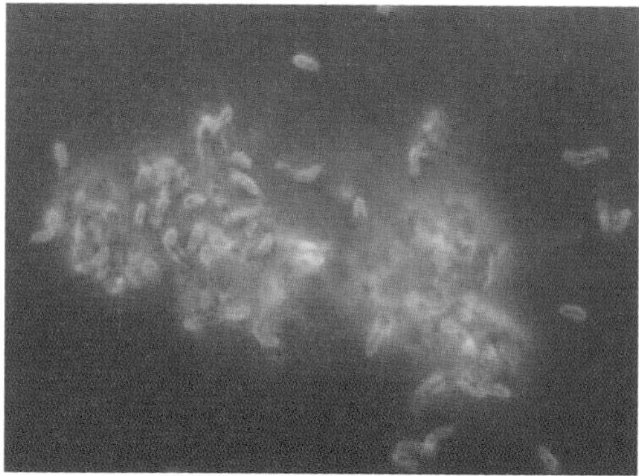

The Science of the Fart. The wonderful Dr. Michael Levitt and the Mylar pantaloons which he invented to investigate the constituents of flatus and role of sulphide-producing bacteria.

I—and some of my team—investigated the subject in great detail with the help of a physician who has spent the last thirty years of his medical career studying fart gas.

Dr. Michael Levitt is, I have to say, one of the nicest and warmest physicians I met while working on this book—he absolutely twinkles and glitters with enthusiasm, interest, and wit. His research into the origins of the word, by the way, have led him to the conclusion that "fart" is not an expletive or a curse word or slang or even vernacular—it's the correct word for "nether flatulence" and is the appropriate way to avoid euphemisms such as "cutting the cheese," "letting one go," "tooting," "dropping one," "tearing the sheet," "letting rip," "breaking wind," "blowing off," "bracing for a warm front," "ventilating the underwear," "testing the posterior wind-tunnel," "blowing the arse-trumpet," or, as they say in Cockney rhyming slang, "horse-and-carting." And to complete the lexicographical survey, the correct spelling of the standard Bronx cheer is PBFLLT. (My own slang dictionaries did yield the unexpected information that a nineteenth-century word for a footman or servant who is forced to follow closely behind his master was "fart-catcher." I think the word disappeared when they attempted to unionize the job in the 1920s.)

At any rate, Dr. Levitt got into farts when he was a junior resident at the Massachusetts General Hospital. By his own account he didn't shine or excel in any particular field (although in my opinion he would have done just as brilliantly in almost any area) and his chief—who took pride in assigning his juniors to various fields of endeavour—stared at him blankly and in some frustration said, "You ...you ...why don't you concentrate on fart gas." To this day, Dr. Levitt is not sure whether this was career advice or a polite curse. Anyway, that's what he did and has made a brilliant success of it—in terms of scientific discoveries and genuinely important research. Just to establish the man's credentials, quite early in his career he discovered that when the body cannot metabolize lactose normally (for example, if there is deficiency of the enzyme *lactase*), bacteria produces hydrogen gas in the colon and this can be the basis of a clever diagnostic test. In fact, Dr. Levitt's hydrogen breath test for lactase deficiency is still in current use all over the world.

Fart Fact

The average person farts eleven times a day. (The upper limit of normal is twenty-three.)

However, we were in his laboratory to get the low-down on farts. And the question we all wanted answered was this: How often does the average person fart in a day? Our research team held a sweepstake and I'm afraid nobody was even close. The correct answer is eleven. An average human farts eleven times a day. (I know that we can all exceed that number in a morning when we're on our own, but this is an *average* number—the upper limit of normal is apparently twenty-three.) Furthermore, the people who suspect that they are producers of abnormal amounts of fart gas are mistaken—most healthy individuals produce the same amount, but some people are more conscious of it and more sensitive to it than others. They're not fart-factories, they only think they are.

Levitt had to devise methods of collecting human fart-gas, and this wasn't as easy as it might seem. In fact, his wife is a seamstress of considerable repute and she designed some purpose-built pantaloons

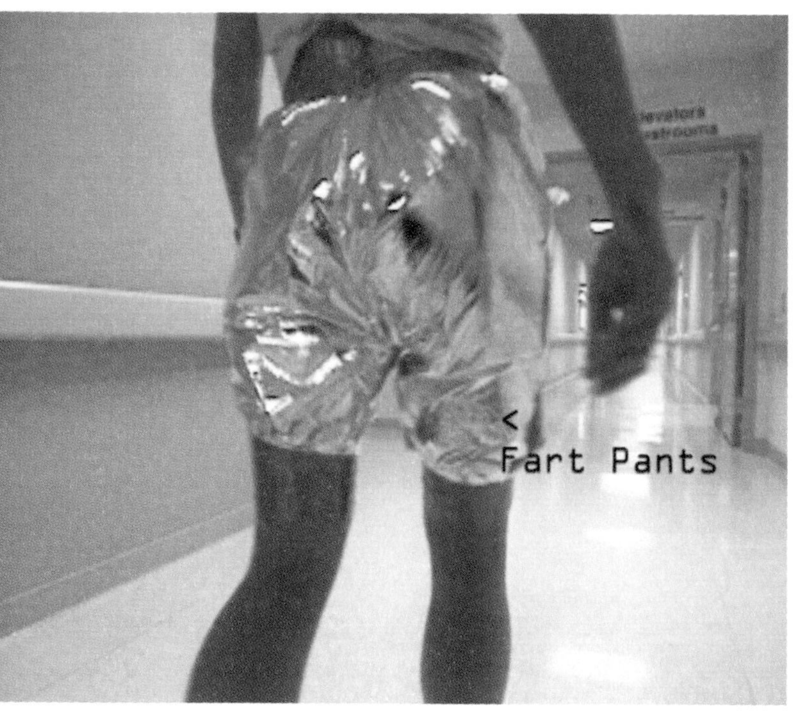

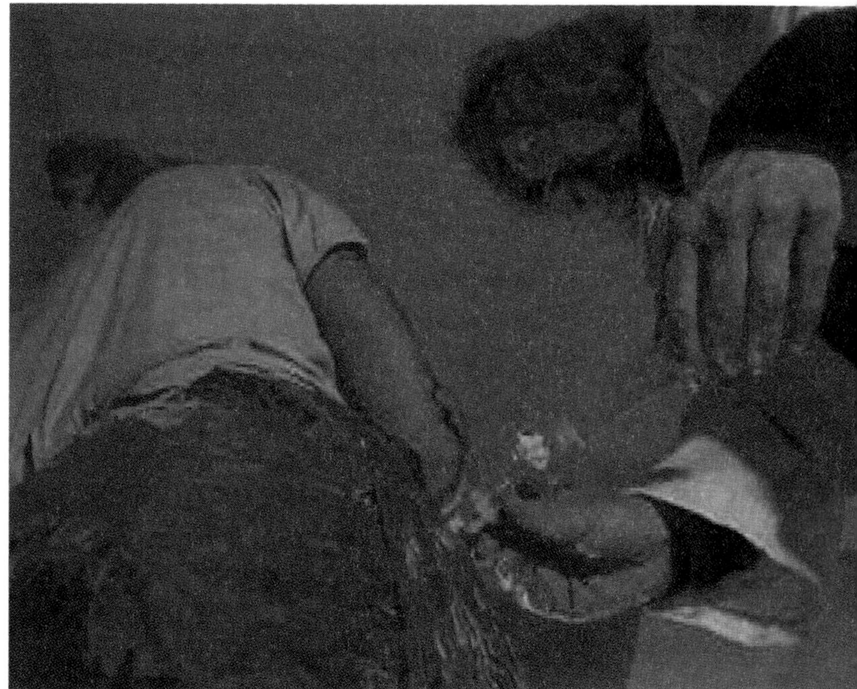

It's an Ill Wind that Blows. . . A volunteer courageously submits to being taped into some unflatteringly tailored Mylar pantaloons. She then walks around courageously while trying courageously to produce flatus, before courageously allowing Dr. Levitt's team to sample the effluent gas and courageously sniff it.

made of Mylar (the gas-resistant fabric used for making hot-air balloons—a singularly appropriate resonance in my view). The pantaloons are made air-tight by being taped to the thighs and round the waist with duct tape (the emblem of true research). The volunteer then walks around for a few hours gradually inflating the Mylar pantaloons with flatulated gas, which is then sampled and analyzed by a chromatograph.

Levitt's research shows that the gas consists mostly of three components: hydrogen, carbon dioxide, and methane (all of which are produced by the activity of colonic bacteria within the bowel), along with nitrogen

Fact—Not Urban Legend

Some farts can be ignited (due to the presence of methane—produced by the decomposition of vegetable matter).

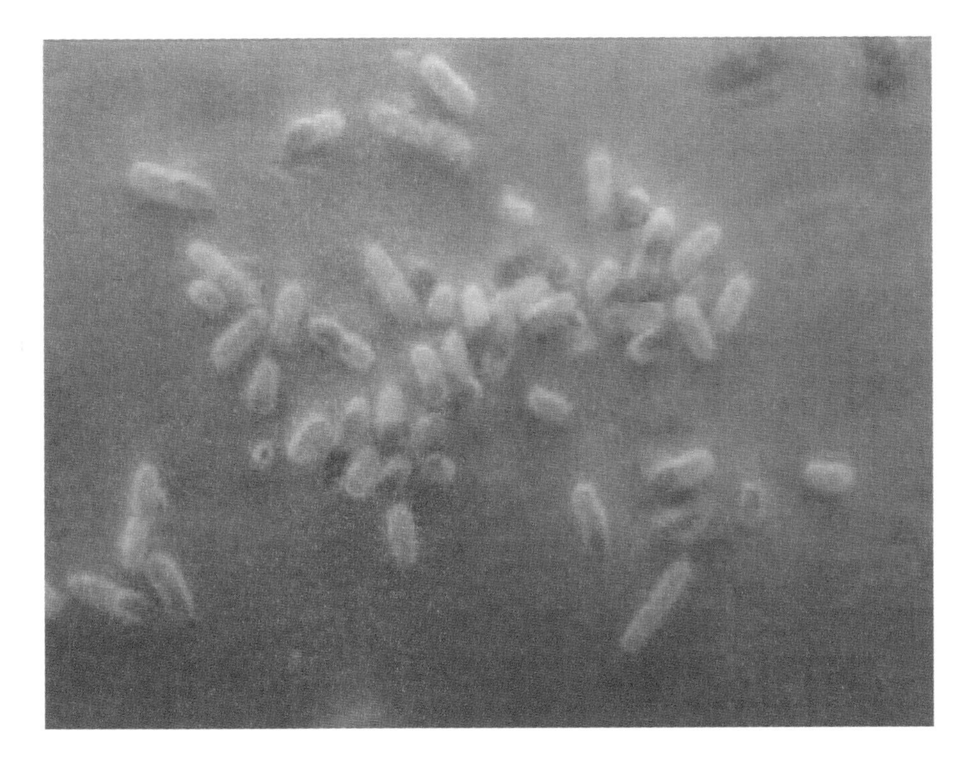

Innocuous but Odiferous. Sulphide-producing bacteria don't look very different from any others, but if nature was fair they'd be decorarted like a skunk.

and some oxygen (both of which are remnants of the air you swallow as you eat). Of these five main components, perhaps the methane deserves an extra mention. Methane is a colourless and odorless gas that is inflammable. So if you were wondering whether farts can genuinely be lit, or whether it is simply one of those urban legends, the answer is yes, farts can be lit and they do burn with a blue flame. In fact, there is a fraternity at a college in the Midwest where the initiation rite is lighting a fart, and the frat members are known colloquially as "The Blue Angels."

However, none of those five gases in fart gas—hydrogen, nitrogen, carbon dioxide, oxygen, or methane—has any appreciable odour. So why do some farts have that characteristic smell? And the answer is that, once again, it is the wildlife residents of the colon that make us repulsive. Within the bowel, there are groups of bacteria that have the capacity to metabolize sulphur-containing amino acids (that is, the capacity to

The Courageous Sniff Tester. Whether certifiable, very brave, or willing to do anything for money, we don't know. But this man and his nose sample the end product of the mylar fart pants.

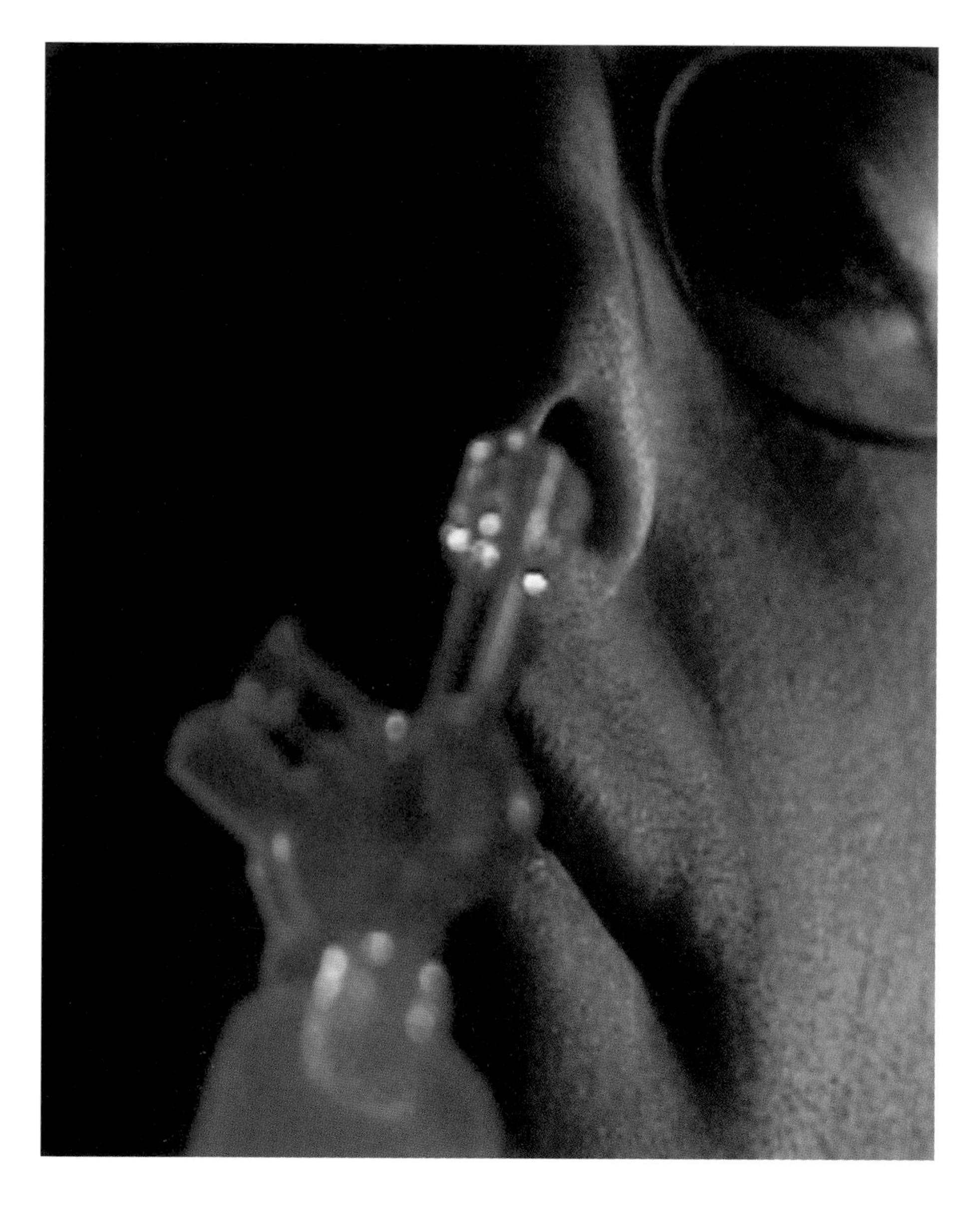

break down the products of certain proteins, particularly those found in meat as opposed to vegetable matter). It is the activity of those sulphur-metabolizing bacteria that produces a range of volatile sulphur compounds (VSCs), and it is those compounds that give the gas its smell. So although vegetarians produce lots of gas, it generally smells less intense, whereas people who eat sulphur-containing proteins (meat) will have smellier gas.

One aspect of this research particularly demonstrated Levitt's persistence and doggedness. A chromatograph can measure which compounds are present in the gas even when they are there in minute proportions—but that analysis alone cannot tell you whether or not the smell is offensive. To determine that, Levitt actually employed a panel of volunteer judges. They took samples of the collected gas in syringes and then courageously inhaled as the syringe was emptied, marking down their opinions as to the level of offensiveness on a questionnaire. It was like a wine-tasting for fart gas. As Dr. Levitt pointed out, you have to pay the people who produced the gas about $70 an hour, but you had to pay the judges much more. Which is as it should be.

Of course there are social elements of farting that cannot be helped by Dr. Levitt's work. Perhaps the most perplexing is what to do if you inadvertently fart while working in a library with other people sitting around you at the table. This farting-in-the-library problem worried me for years, until I invented the answer. You simply raise your head and glare at the person sitting next to you. You know it wasn't him, but—colonic activity being just a tad unpredictable—he won't be quite as certain. With a bit of luck, he'll be embarrassed and unsure of himself and will look guilty even though he isn't. If you're really in luck—and I've only achieved this once—he'll be so embarrassed and flustered that he actually will fart. In which case you sadly shake your head and settle back to your work, with an inner glow and warmth that is entirely unrelated to gastrointestinal bacteria.

On a more serious note, I thoroughly enjoyed our day with Dr. Levitt. The depth of his knowledge and his track record in research, combined with his philosophical view of what really matters in life, impressed me greatly. The achievements of Dr. Levitt's career prove, in my opinion, the old saying that truth is like daylight shining behind a curtain with many holes in it—how much you see does not depend on which hole you look through, but on how close to the curtain you can get your eye.

The Origin of Feces

The future of human civilization will always depend on the successful separation of sewage and drinking water.

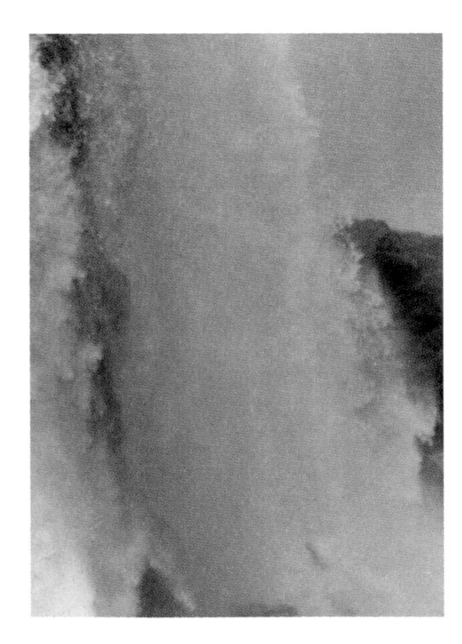

Forgive the slightly specious pun on Charles Darwin's *magnum opus*, but actually this chapter is going to be fairly grown-up about an awkward and embarrassing aspect of human behaviour—defecating. Moreover, it will show you that in a very important way, our future as a species depends on how and where we dispose of our sewage.

OHHH! YECCH! PYEW!

We all know—without being told—what stool smells like, and nearly all of us acquire a serious aversion to the smell and appearance of stool at a very early age (usually after an initial fascination that is quite normal). If we ever think about it at all, we generally imagine that stool has a "disgusting" aroma naturally, and that we are appropriately programmed to move away from things that smell and look disgusting. In fact, it might be the other way round—we might perhaps be defining the state of "disgusting" based on the smell and appearance of stool because it is highly likely to represent a health hazard. It may not be a case of "stool is disgusting" but more a case of "we call it disgusting if it resembles stool."

This distinction may sound trivial, but it probably isn't—and it may underlie and influence all of our social attitudes to matters of excreta. Our instinctual feelings of revulsion and the urge to back away and withdraw from anything that looks and/or smells like stool may be a sophisticated and socialized remnant—so some behaviouralists believe—of our abreaction to feces. The behavioural pattern "if-it-smells-like-stool-move-away-from-it" presumably proved its worth early in evolution, since animals that did not move away

Things I Didn't Know (and Didn't Really Want to)

The average person, during a single day, deposits in his or her underwear an amount of fecal bacteria equal to the weight of a quarter of a peanut.

were more likely to be poisoned. To us, feces smell dreadful because we have a built-in association that links a particular type of smell to a health hazard—in other words, our definition of "terrible" as regards a smell is a marker of our ability to link that kind of smell with danger. It may be the biological underpinning of the withdrawal response that gives the strength and the force to the feelings of "Yecch!"

Let's start at the very beginning. When there were only a few members of our species in existence, perhaps a hundred thousand or so, problems with disease transmission were relatively few and fairly simple. There may have been the occasional outbreak of pneumonia among the huddled not-yet-masses, but even then, the weaker specimens would succumb and the genetics of the stronger types would get a still-greater advantage from the skewed wheel of natural selection. A couple of million years later and the story is very different. In modern times, the infrastructures of our vast cities are so complex and so intricate that what happens to one person can very quickly become a neighbourhood issue and then—with mass rapid travel being so accessible—a national and then an international problem. And much of it starts with sewage.

Problems with Our Oral Traditions

Whatever the font of our feelings of revulsion, we can look relatively dispassionately at the biological hazards of fecal material. At this stage in humankind's history, it really matters.

A Global Perspective

If you were able to look at our planet from outside, you would instantly be struck by the fact that what are severe health problems in many areas of the world have been almost obliterated in others. To be more specific, the infectious illnesses that kill the largest number of young people—diarrheal illness—cause relatively few deaths in the developed world. In the developing countries, however, the toll is terrible (a message that is steadily getting through). In itself that is tragic and unfair, but what makes it worse is that the remedy is not—in global terms—very expensive.

It is estimated by the World Health Organization (WHO) that, at this moment, more than one billion people do not have access to drinkable water, and about 2.4 billion people live in areas without adequate sanitation. They estimate that the cost of fixing both of these things—provid-

Disturbing Global Facts

1. One billion people do not have access to drinkable water.
2. About 2.4 billion people live in areas without adequate sanitation.
3. The remedy for both problems would cost approximately US$10 billion per year.
4. US$10 billion is half the amount spent in the US on pet food per year and is about the same that is spent in Europe on ice cream each year.

ing the majority of people on earth with clean water and sanitation—would be approximately US$10 billion per year. And as Dr. Philip Tierno points out in his book *The Secret Life of Germs,* this sum—$10 billion—is half of the amount spent by US consumers on pet food and about the same amount that Europeans spend each year on ice cream.

On a worldwide scale, then, the transmission of disease by the contamination of drinking water by sewage is clearly a very serious and very basic problem. Yet even when "the plumbing has been fixed," there are still very important hygienic practices that need to be observed—as we shall see. People who are fortunate enough to live in places where they can get clean drinking water and do have sanitation, still need to know a bit about the potential dangers and about what they should be doing to avoid them.

The bottom line is actually fairly simple: when your mother told you to wash your hands before eating, she was absolutely right. The simple fact of the matter is that we humans can harbour and transport many kinds of nasty species of wildlife in our colons (and associated areas). By the time you've read this chapter, you will have a much better idea of what precautions are sensible and should be followed regularly—and which ones are merely faddy or perhaps obsessive. The secret to staying healthy begins with knowing how and when to wash your hands.

Intriguing Thought of the Week

"The most heavily contaminated area in the washroom is the underside of the toilet lid. So try never to lift the toilet seat yourself—have someone else do it for you."

Dr. Chuck Gerba, microbiologist, University of Arizona

The Toilet Seat of All Learning

Of all the scientists that I interviewed for this book, Dr. Charles Gerba was the one who looked most like the Hollywood stereotype of the nutty professor. He is an extraordinary man in many respects: first, he moves and talks at only slightly less than the speed of light, secondly, he knows virtually everything about the microbiology of human habitats, and thirdly, he wears that learning lightly. As a domestic microbiologist at the University of Arizona, he has spent most of his recent career perfecting techniques for diagnosing the presence of the type of bacteria found in the bowel (and therefore in stool)—*E. coli* and like-minded spirits—all over the house.

Dr. Gerba's instant testing system for these types of bacteria is quite sophisticated but relatively simple to use. A cotton swab is taken of the area of interest—the kitchen counter, the toilet seat, the toothbrush holder, the washing machine, or whatever—and the swab is put into a special solution that converts the presence of any of the colonic-type bacteria into a chemical reaction that produces light. The amount of light emitted—as measured by an instrument called a *luminometer*—is directly proportional to the number of bacteria on the swab. Hence the luminometer can tell you (approximately) how heavily contaminated the surface is.

The instant read-out given by the luminometer is extremely useful, but it does not necessarily reflect the proportion of *E. coli* bacteria among the contaminants. In order to accurately assess the amount of *E. coli*—which will give a more accurate indication of fecal contamination because they are the predominant species of bacteria in our stool—Gerba has developed a much more specific test. The swab is immersed in a solution that converts specific antibody reaction against *E. coli* into a yellow colour. After overnight incubation, the various samples are evaluated; the deeper the colour, the larger the number of *E. coli* present. In fact, if the *E. coli* contamination is really heavy, the contents of the sample tube will fluoresce.

Upstairs and Downstairs and in Milady's Chamber

Gerba has taken literally tens of thousands of samples from almost every domestic and commercial setting that you can imagine (and several that you can't) and his results are extremely surprising—and a little disturbing until you hear him explain the significance of them.

For a start, in the average household, there are fecal bacteria everywhere. You might think that the toilet would be the most heavily contaminated area, but that's not actually true. The top of the toilet seat has virtually no fecal bacteria on it. On the underside of the toilet, however, the contamination was very heavy. As Dr. Gerba explained, this is not because the users are inaccurate or perverse in their aim when excreting.

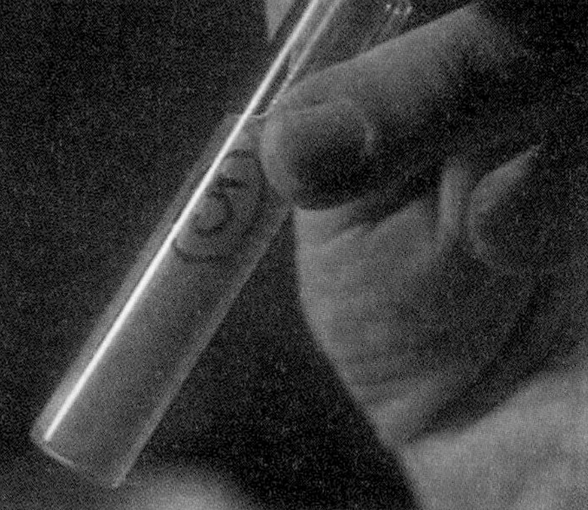

When the Fan Tracks the Sh*t. The wonderful and ever-active Dr. Charles Gerba traces the presence of fecal bacteria in every setting in the home.

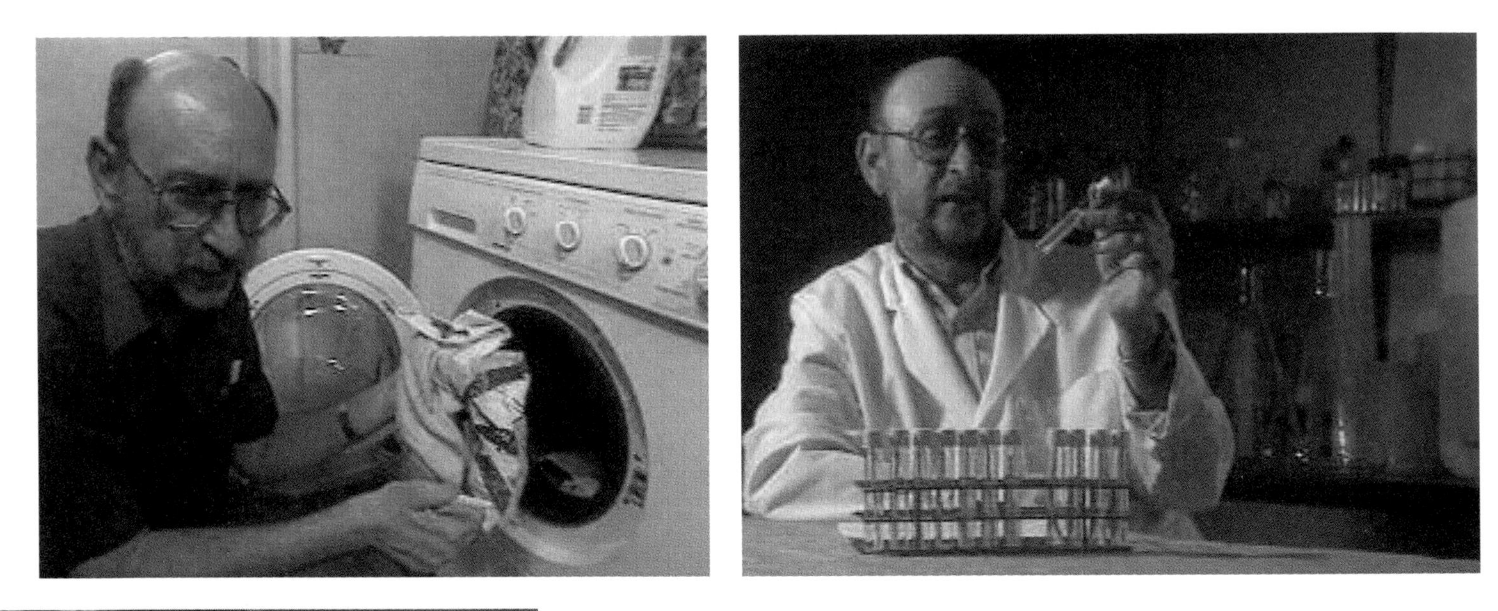

The reason is actually because the water flushing down the bowl creates a very powerful aerosol of droplets that scatter—literally—far and wide. In an average home, when the lid is open, droplets may travel up to twenty feet from the bowl—certainly as far as the average distance separating the toilet from the toothbrush. In a public toilet where the flushing mechanism is much stronger, the aerosol is heavier and travels further. When the lid is closed the spray travels less far, but—as the tests show—the underside of the lid is heavily contaminated.

In the kitchen, however, the problems are more severe. The tests show that—with fecal bacteria derived from food—the kitchen sink is the most heavily contaminated site in the house. "If I were a microbiologist from Mars," said Dr. Gerba, "I'd look at these test results and I'd use the kitchen sink as my toilet and eat my dinner on the toilet-seat." Even heavier contamination occurs on the kitchen sponge, When you wipe down the counter with a kitchen sponge, Gerba explains, you are giving the bacteria a free ride around the kitchen. Contamination in houses where the person cleans the counter regularly are very heavy, whereas contamination in homes run by bachelors are lighter because (according to Gerba) bachelors don't clean.

I wasn't very reassured by what Dr. Gerba said about the chopping board, either. According to his tests, the average plastic chopping board has got more bacteria on it than the top of the toilet seat. Personally, I am not yet ready to start preparing my salads on the toilet seat, but even so, we should all pay heed to his results.

Things Nobody Can Bear to Think About

The top of the toilet seat is contaminated with fewer bacteria than the average chopping board.

***E. coli* Stained Red. As Common as Muck.** These are the commonest bacteria in human stool. (The photograph has been tinted red. In real life, the colour is somewhat different.)

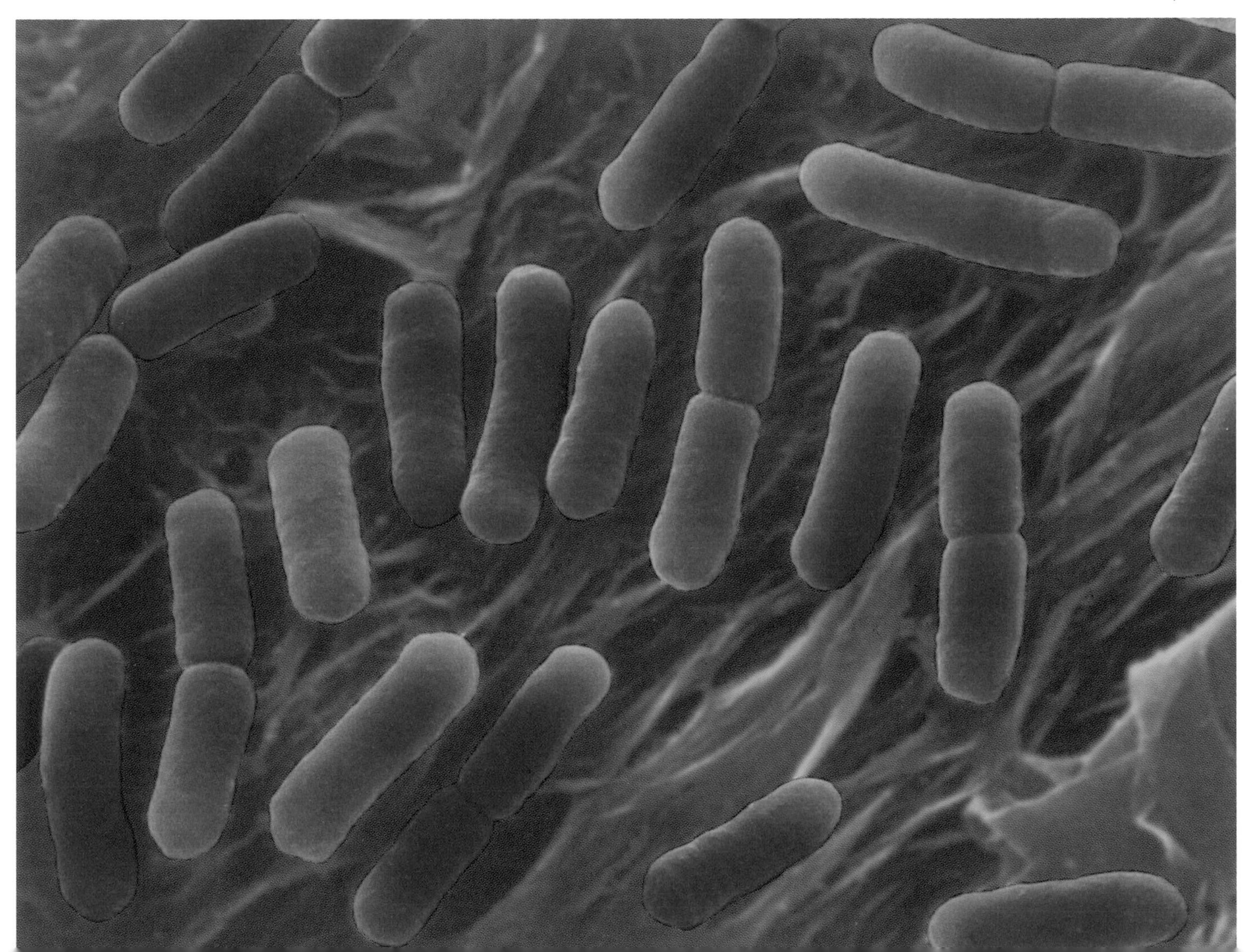

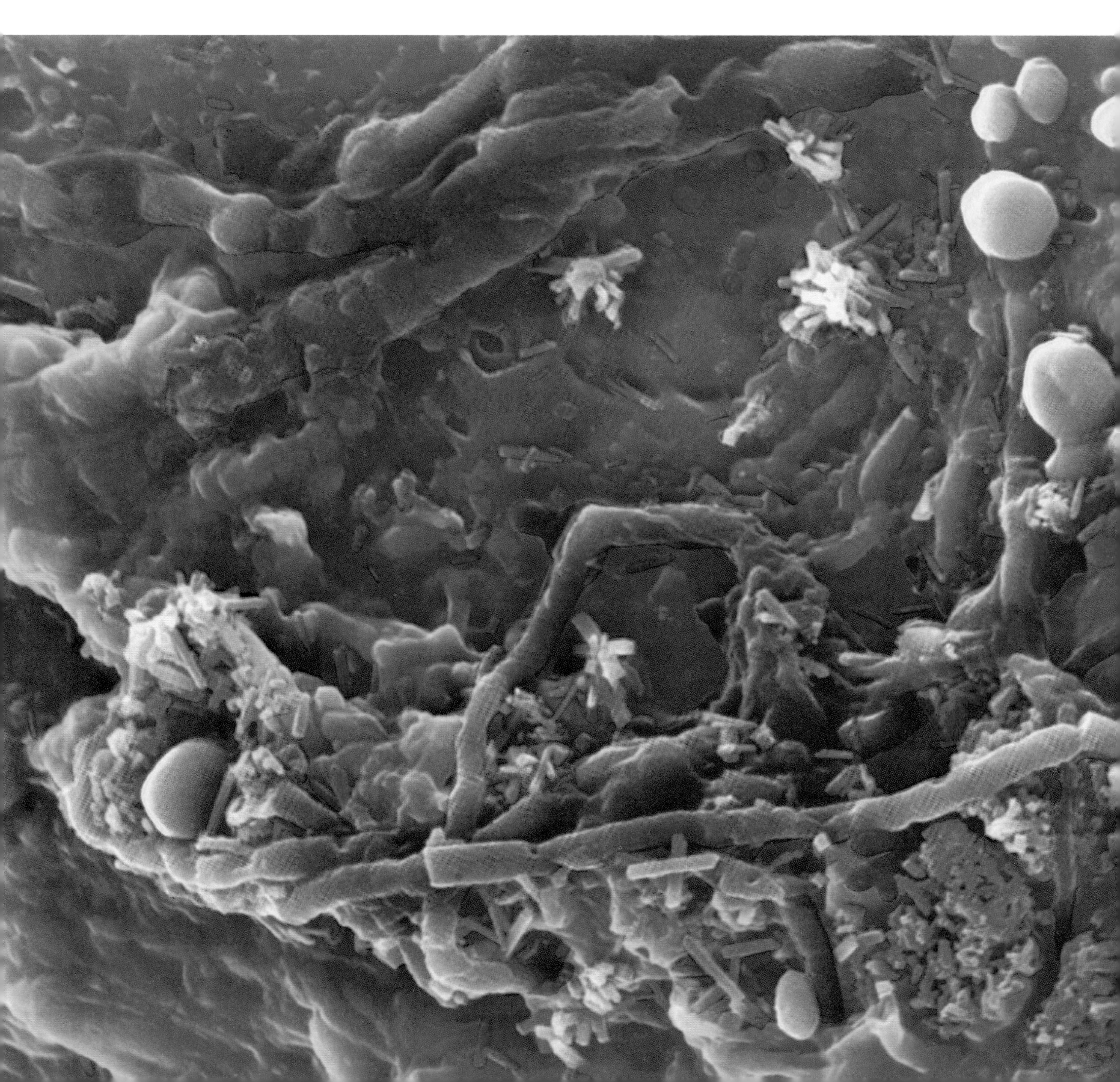

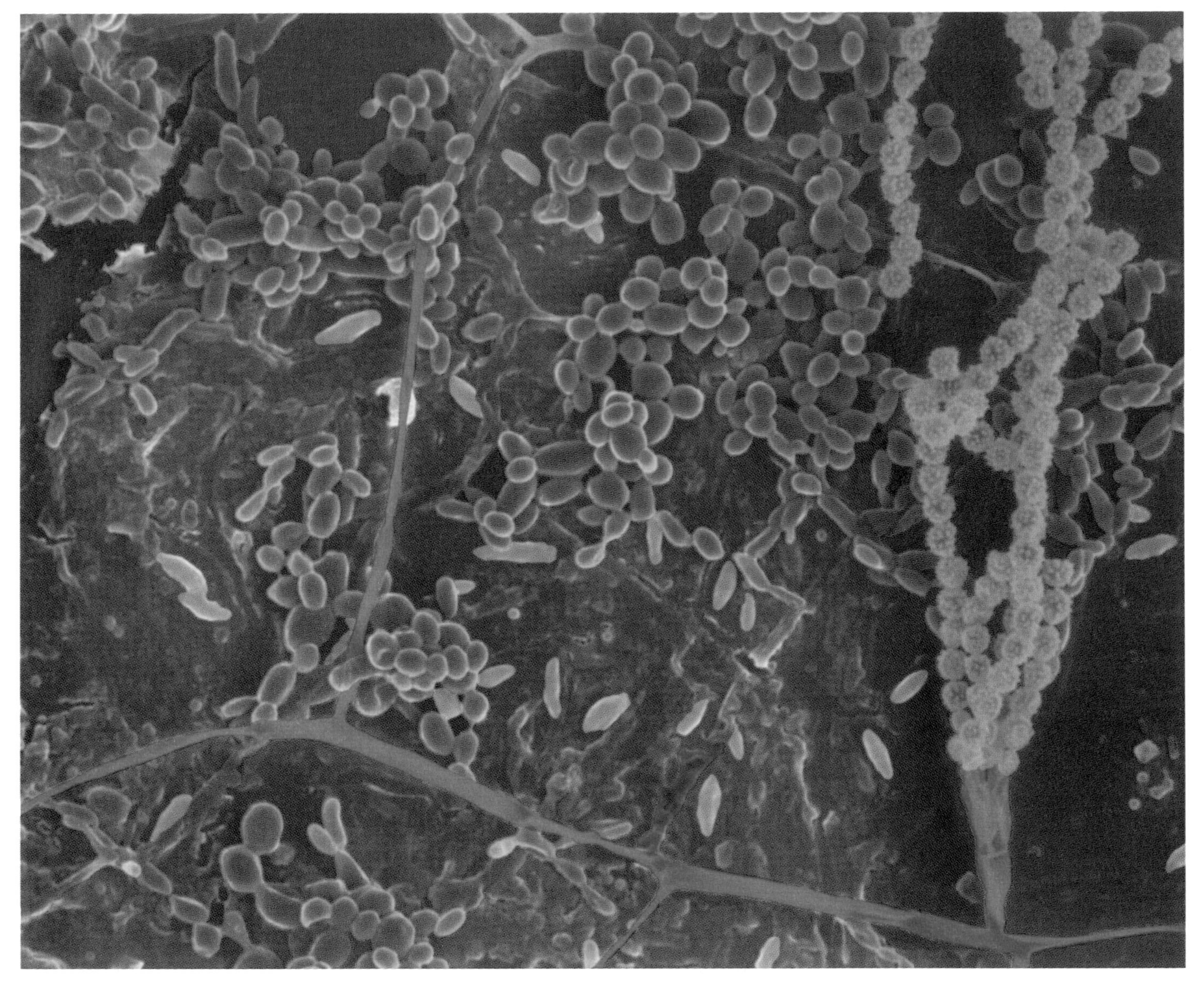

(Opposite)
Wildlife Hitches a Ride. This microscope picture shows the wildlife—bacteria (rods are purple, cocci are blue) and fungi (branches are red and yeasts are yellow) that live on an ordinary kitchen sponge. As Dr. Gerba says, when you clean with this, you are just giving the bugs a free ride.

The Chairman of the Board. These are the resident bacteria and fungi on the average chopping board. In this picture the rods (bacilli) are green, the cocci are brown, and the fungal branches and spores are orange and pink.

The second most heavily contaminated place in the house is the laundry room. The reasons are perhaps obvious: underwear is—not unexpectedly—teeming with fecal bacteria. "The average person's underwear contains at the end of a day," states Dr. Gerba, "an amount of fecal bacteria equal to the weight of a quarter of a peanut." I'm not sure that we actually have to specify the airline that provided the peanut he used for comparison—I think the principle is clear anyway. When he did his tests on the trap—where the waste water drains—inside the washing machine, the level of fecal contamination was second only to that of the kitchen sink. The inside of the washing machine is moist and warm most of the time, so it is to be expected that contamination would be heavy. Gerba's point in all this is that after handling the laundry, you should wash your hands very carefully before preparing food or eating.

Arcane Trivia for the Anxious Obsessive

The middle stall in a row of toilets is usually the one that has the heaviest bacterial contamination. If you really worry about that, use the stall nearest the door.

So What?

The really crucial question in all this is, Does any of it matter? Does it matter that your toothbrush in your bathroom is getting regularly sprayed with fecal bacteria from your toilet? Does it matter that your kitchen sink is almost lavatorial in its microbiology? The answer is that it probably doesn't matter very much as long as you know about it. Dr. Gerba's advice is phrased frivolously, but the basic message is an excellent one—know where the bugs are. Be aware that the kitchen sink is an absolute swamp of bacteria—many of them of fecal origin. That doesn't matter so much in itself, but it certainly matters if you drop food in the sink that you want to eat without cooking or peeling. One would almost be tempted to say—in Chuck Gerba's addictive style—"If it falls on the toilet seat, you can think about washing it and eating it, but if it falls in the kitchen sink and you can't cook it or peel it, think about throwing it away." We don't need to be paranoid and we don't need to live in a sterile bubble, frightened of touching anything—but we do need to be well informed about where the bugs are. Above all, we need—as we shall see later—to wash our hands before touching food and eating.

Things I Didn't Even Dream of Knowing

Freshly washed laundry is teeming with bacteria—don't prepare food or eat after handling the laundry until you've washed your hands really thoroughly. (No matter how clean you THINK you are!)

Have a look at the information box for some more practical tips—ALL of which were new to me until I started this book. (Although if I had listened to my wife—as I now do—I would have been working with these rules for many years by now.)

If you can, close the lid of your toilet when flushing (or have someone else do it for you), but don't worry too much about your toothbrush or facecloth—they've been getting toilet water sprays for years and it hasn't done you any real harm.

Simple Practical Tips for Food Handling in the Kitchen

1. If you drop a piece of food into the kitchen sink, don't even think of eating it unless you can clean it very thoroughly—or peel it or cook it.
2. Take real care of your chopping boards—have two of them, one for salads and vegetables and one for meat, and never chop vegetables on the meat one. Clean them with a scrubbing brush after use, wipe with a diluted solution of bleach, and dry.
3. When you get chicken or beef into your kitchen, handle it as little as possible before cooking—the less contact it makes with your work surfaces the better.
4. Always cook chicken thoroughly—and refrigerate it properly after cooking if you have to.
5. If you are cooking steaks or cuts of beef or lamb, it's all right to cook the outside and leave the inside rare if you wish, but if it's ground beef (so that you don't know which is inside and which is outside), cook it all very thoroughly.

Snow Days

It has often been said that the sign of a really major revolution in thinking—be it in art or in science—is that after it has happened nobody can imagine what life was like before it, or why the revolution caused so much fuss. A great revolution supercedes itself, as the saying goes. It is very difficult for us to imagine the concept of the stars and the heavens before Galileo (and Copernicus and Kepler) suggested that the sun stays still and the earth moves round it (despite the evidence from our eyes every day that this is visibly nonsensical). A couple of generations ago the world of literature was shaken to the core by the prose style of Ernest Hemingway—now we can hardly imagine why it was so shocking. The same is true in imagining evolution before Charles Darwin: most of us—there are still one or two diehards—can only think of the life forms on this planet as having evolved over billions of years, not created in a single week approximately ten thousand years ago.

So it is with the theory of germs and infections. It is difficult to imagine what it was like before Louis Pasteur, before Ignaz Semmelweis, and before Robert Koch. Those three people—and there were many others who didn't make it into the Bacteriological Hall of Fame—suggested that diseases were caused by things, things so small you couldn't see them. Before that—and this is where it gets really difficult to believe—people thought that diseases just *were.* People got cholera because it just happened. Tuberculosis was caused by having an artistic temperament and being a sensitive person—the type of personality was actually called *phthisical* from *phthisis,* an antique word for TB. Malaria, so people thought, was caused by "bad air" from swamps.

In fact, outbreaks of cholera were so common in towns and cities up to the end of the nineteenth century that they were almost regarded as a fact of life. Until a physician called Dr. John Snow came along. In London, cholera outbreaks were disastrous but occurred very often; in 1854 alone the number of deaths from cholera in Britain was close to ten thousand. Through painstaking investigation into the causes and transmission of the disease, John Snow began to suspect cholera outbreaks had something to do with water. (The idea of water carrying diseases wasn't entirely new—apparently Hippocrates in the fifth century BCE wrote about it and Alexander the Great planned some of his camp sites to separate drinking water from sewage. But if that knowledge had ever been widely disseminated, it had certainly been forgotten in nineteenth-century London.)

Dr. Snow noticed that the number of cases of cholera differed markedly between areas that were supplied with water from different sources. For example, in the Southwark area of London, where the water supply was river water contaminated with sewage, there were 315 deaths from cholera per 10,000 houses. By contrast, Lambeth, which had a different and cleaner water supply, had far fewer deaths—37 per 10,000 houses. Furthermore, the extraordinary Dr. Snow actually carried out an experiment in which he took a bottle of water from a cholera area and gave it to a widow and her daughter living in an area where there was no cholera. Both mother and daughter developed cholera. (Nowadays, that experiment would be regarded as unethical—well, criminal, actually.)

In any event, in 1854 there was a serous outbreak of cholera in an area of London called Soho (named after a rallying cry of fox hunters "So Ho!"). Dr. Snow realized that almost all of the people who got cholera were

Handling Cholera. Or rather unhandling it. This is a replica of the public pump in Broad Street that was the source of a cholera outbreak. Dr. John Snow, whose eponymous pub is in the background, unscrewed its handle, rendering the pump useless and the local citizenry safe.

Anne Hardy. put it beautifully when she described the problems of nineteenth-century London's contamination of water by sewage as "leakage and soakage."

using one particular pump in a street called Broad Street (nowadays named Broadwick Street). With the single-minded grit and determination that comes only with a true sense of a public-health mission, John Snow went up to the Broad Street pump and unscrewed the pumphandle.

The people couldn't get their water and there was a furious public outcry, but as the old English phrase has it, "the proof of the pudding was in the eating." When the fuss died down, the cholera outbreak had died out. Dr. Snow had proven that contaminated water was the source of cholera. At the time about 10 per cent of Londoners got their water supply from big public wells, and sewage disposal was rudimentary (that is to say, absent—everyone just dug their own cesspit). By the end of the century, nearly 90 per cent of Londoners enjoyed a civilized separation of their sewage and drinking water. They were—as Dr. Anne Hardy of the Wellcome Trust Centre for the History of Medicine at University College London put it—no longer prey to diseases caused by the mingling of those two things; there was very little "leakage and soakage." As for John Snow, he did very well, and eventually had a pub named after him. No public health official before or since has been able to claim that honour.

The Fecal-Oral Railway

What Dr. Snow demonstrated that day in Soho was that *something* in the drinking water from a public well was the cause of cholera. Now we know that it was the bacterium—*Vibrio cholerae*—and as you will have guessed from the title of this chapter, cholera is just one example of wildlife that can be passed from one human to another by means of stool.

As I said at the beginning of this chapter, the central problem is really over-population. When this planet held only a few thousand humans (or, rather, hominid apes) the chance of one of them passing on a bug to many others was tiny. There was for the most part so much space that our ancestors could use almost anywhere as a bathroom and nobody else would get into trouble as a result. The chance of feces from Person A getting into the food or drink of Person B was infinitesimally small.

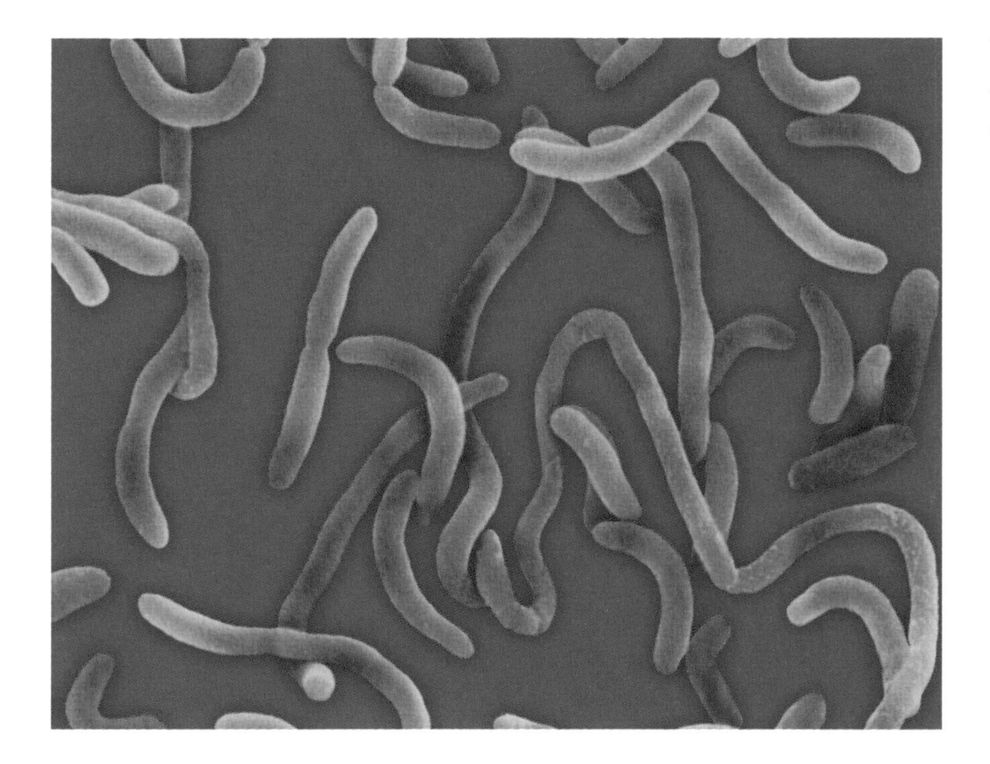

Unwelcome Drinking Partner. This is the *Vibrio cholerae,* the bacterium that causes the horrid and sometimes fatal enteritis known as cholera.

As the human population increased and communities grew up, however, people became concentrated in groups and lived in closer contact with each other. This aspect of human social organization made a major leap about 10,000 years ago when we first started developing crops for food—we started farming. At that point, human societies became, as the saying goes "rooted to the spot." With an (almost) guaranteed food supply, our ancestors lived in the same place year after year, instead of having to wander around being hunters and gatherers. This was undoubtedly a good thing, but it greatly increased the chances of transmission of bacteria and other parasites from one person's stool into another person's drinking water or food. This is what is meant by the phrase "fecal-oral transmission"—and it's a major factor in human disease that we need to know about.

At the time of writing, there are about six and half billion of us on this planet and many of us live in big cities where the drinking water

權記
膳食中心

comes from large community resources and sewage disposal is also a community responsibility. As we saw from the example of cholera in nineteenth-century London, contamination was a very serious and common problem in the growing stages of big cities. When people just dug their own cesspit anywhere they wanted, there would be flooding after heavy rainfall. If the people just dug a new well wherever they wanted—which is what often happens as communities develop—the chance of fecal-oral transmission of infections increases sharply.

(Opposite)
Mixed Blessings. The human species is, on average, a hundred thousand times more populous than any other primate. If this was a marketplace of chimpanzees, there would probably be 0.0002 of them in the photograph.

Fewer of Us. Despite the lack of medical amenities and healthcare, early rural communities did not have many problems from water-borne infections.

(Inset)
Close Neighbours. At low population densities we rarely pollute our neighbour's water, air, or food. Urban development means that we have to take a lot of care to separate our sewage from our drinking water, and food.

The point about fecal-oral transmission is that it can be direct or indirect. Direct transmission occurs when sewage goes straight into the water supply. Transmission can also occur indirectly when sewage ends up on objects such as food—fruit and vegetables are a real problem if they are eaten uncooked—or on inanimate objects that are touched by the infected person and then by the unwitting victim. In other words the problem can exist on both a community scale—when something goes wrong with the water-sewage separation—or an individual level. As you'll see in the examples that follow, this is something that we need to be constantly aware of in designing and regulating our sources of food and water and our disposal of sewage. Anne Hardy is absolutely right—"leakage and soakage" are community friends of the fecal-oral railway and enemies of society.

The Case of the Imported Raspberries

From the point of view of avoiding the fecal-oral transmission of unwanted wildlife, there are two important factors in preparing food: what are the characteristics of the particular food, and where it comes from. As with many examples in this book, the objective is not finger-pointing or assigning blame; it's all about knowledge and what measures can be taken to reduce the chance of problems to an absolute minimum.

The culprit here, called *Cyclospora*, is another life form that can travel with great facility along the fecal-oral railway—in this case using hijacked raspberries as its personal railway car.

The Toronto outbreak in 1998 began after what was undoubtedly—in all respects other than the consequences we are concerned with—a very pleasant dinner party. Unfortunately, several hours later about twenty of the guests—and the hostess—began to get quite severe abdominal problems and diarrhea. A sudden outbreak of diarrhea—particularly in a very defined geographical area—merits an urgent phone call to the local medical officer of health. In the case of Toronto, the person who took the call was Dr. Barbara Yaffe , who runs a very efficient and knowledgeable public health unit.

The stool specimens of the victims had all identified the causative organism as *Cyclospora*—so the problem now became one of finding out which food had been contaminated. The first thing that Dr. Yaffe's unit did was to take very detailed food histories—what foods were eaten at the party, what ingredients went into each dish, and where the various ingredients came from. This is extremely painstaking and hard work, particularly since the individual ingredients have to be traced—where they were bought, where the shop got them from, where the wholesaler bought them, the country that supplied the wholesaler, and so on.

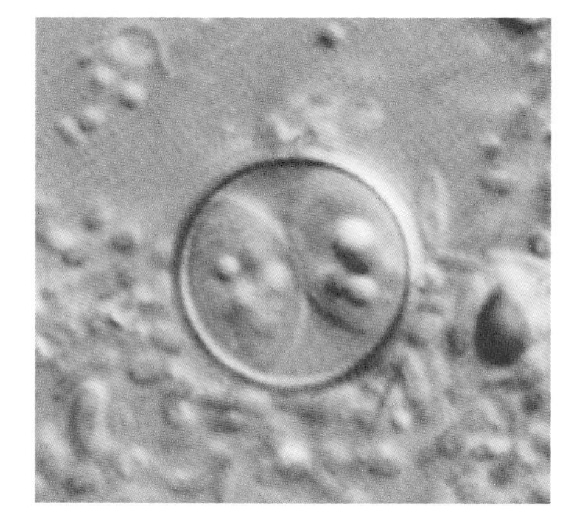

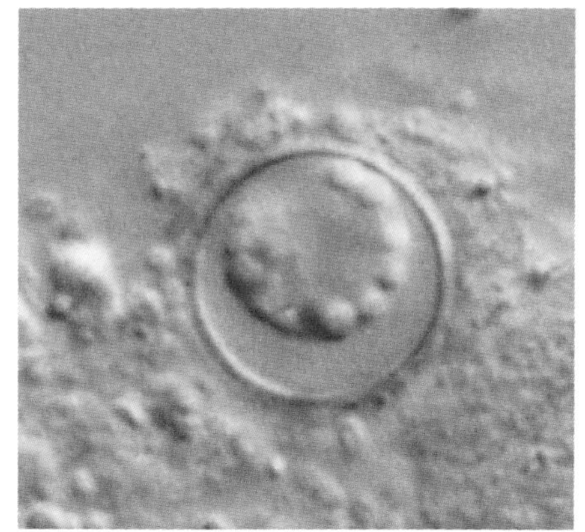

"Undesirable Alien." *Cyclospora* is not a particularly common contaminant on certain fruits—and regulations covering food importation and inspection keep it that way.

Steadily, a picture emerged that pointed towards raspberries as the most likely bearer of the infection—especially likely because raspberries have a very uneven surface with deep crevices and little hairs. It is very easy for any material —including soil or dirt contaminated with feces—to become lodged there. Once the material has hardened around the hair or in the crevice, it can be virtually impossible to remove by washing without destroying the raspberry entirely. Hence raspberries—from the bug's point of view—are very useful as a means of transport along the fecal-oral railway.

In fact, this was still more likely since outbreaks of *Cyclospora* had occurred a few times elsewhere and similar detective work had led to raspberries as the likely vector. In this case, the evidence became clearer and clearer—the raspberries at the dinner party that had been imported from Guatemala were the prime suspects.

So how did the *Cyclospora* get there? The probable answer is that soil contaminated with feces got onto the fruit. This does not necessarily mean that individual fruit pickers were lax in their hygiene; it is also possible that sewage disposal in those areas is not perfect and that

Huge and Hairy. This greatly magnified picture of the surface of a raspberry shows the crevices and little hairs. Dirt can become lodged there and, after it has dried, can be almost impossible to dislodge or even detect with the naked eye. If that dirt contains *Cyclospora* and you eat the berries, you'll have new wildlife inside you.

sewage contaminated the soil. The exact details are still not absolutely clear, and there is a remote possibility that the *Cyclospora* was brought in by birds. Bird feces are certainly found on the raspberry vines, but on average there are more pieces of evidence suggesting human agency than avian.

In any event, the importation of Guatemalan raspberries was halted for the time being. A greatly enhanced testing program has been introduced in which every single farm in Guatemala has to submit to rigorous and regular testing. At the time of writing the situation has improved greatly, and many farms have been allowed to renew the export of their raspberries to certain countries.

The point of this example is very clear. There is no way that the people at the party in Toronto could possibly have known in advance that the raspberries carried *Cyclospora*. Outbreaks like this do happen from time to time—I would almost suggest that occasional and limited outbreaks are

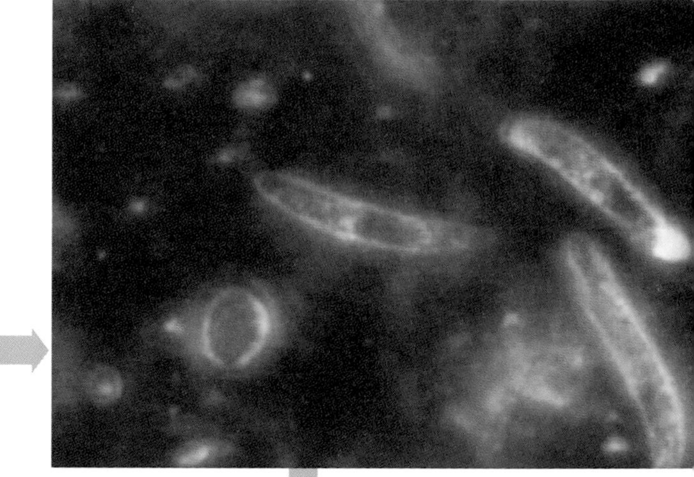

Doctor Detective. It was Dr. Barbara Jaffe's team that did the detective work connecting the Toronto outbreak with some raspberries imported from Guatemala.

virtually unpreventable. It would be utterly impossible for every shopper to read every ingredient of every single food you buy and then ask the supermarket to identify the country of origin for every singe ingredient. But the lessons of the Toronto *Cyclospora* outbreak are actually reassuring—the system worked. With diligent work the public health authorities were able to trace the source and take the appropriate steps to, first, stop any further spread of the outbreak, and, secondly, to prevent—as far as possible—it happening again in the future. The whole event was very unpleasant for everyone—the party guests, the health departments, the farmers, and the government organizations—but as a result, improvements have already been made. It's a bit of a silver lining, at the very least.

The Milwaukee Outbreak

In the spring of 1993, Milwaukee suffered the largest water-borne infection ever recorded. Within a few days, dozens, and then hundreds, of people developed sudden and severe diarrhea. Frank Wilson, the pilot of our boat tour on Lake Michigan, was a victim of that particular outbreak and, as he described it, the diarrhea was so severe and so persistent that as soon as you got up from the toilet, you immediately had to sit down again. You ended up with a sore butt simply from sitting all the time.

The outbreak became more and more severe and widespread, involving thousands and then tens of thousands of people. One of the central microbiologists involved was Dr. Dean Cliver (then of the University of Wisconsin in Madison), who kindly agreed to revisit Milwaukee with me and relive some of his memories of the event. Soon after the outbreak started, Dr. Cliver remembers, the stool specimens of the victims had all shown that the causative bug was a nasty protozoan parasite called *Cryptosporidium*. What makes it so particularly nasty is that it hangs around in a kind of shell—called an *oocyst*—that is unfortunately quite resistant to the normal disinfecting processes used in water purification, and is even partly resistant to chlorine.

There were many factors involved—and there was no single fault or blame that could be laid at any one official's door. The amount of waste coming in from the plumbing of the city and the environs varied, but the significant issue was that the city's system did not have the capacity to accommodate both urban waste water and the run-off water from the agricultural lands around the city. Unfortunately Milwaukee is situated in an area where the volumes of water in the spring runoff can vary dramatically and in 1993 the runoff was particularly heavy.

The system simply couldn't cope with the increased waste from the growing population and the greatly increased influx from the runoff. Also, the inlet from Lake Michigan was too close to the sewage effluent, which meant that the purification plants were getting water that was too heavily contaminated—regardless of whether the source of contamination

Casting a Spell. Dr. Dean Cliver said modestly that he was selected as a lead microbiologist in the Milwaukee incident because he was one of the few people who could spell *Cryptosporidium* without looking it up.

Water Foe. These are *Cryptosporidium* organisms on the wall of an intestine. They are the bugs that caused the Milwaukee outbreak in 1993, the largest outbreak of water-borne infection ever recorded.

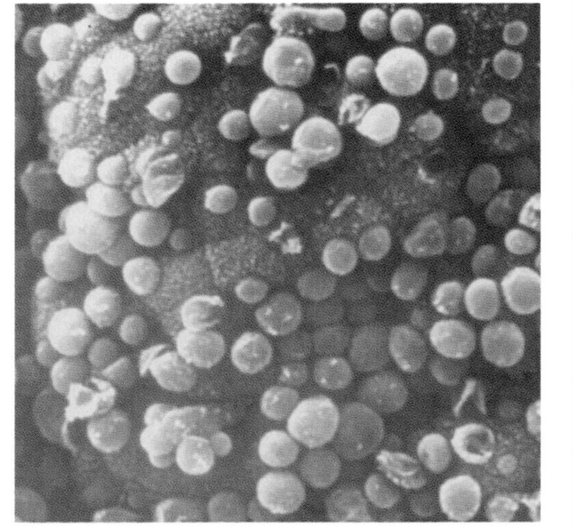

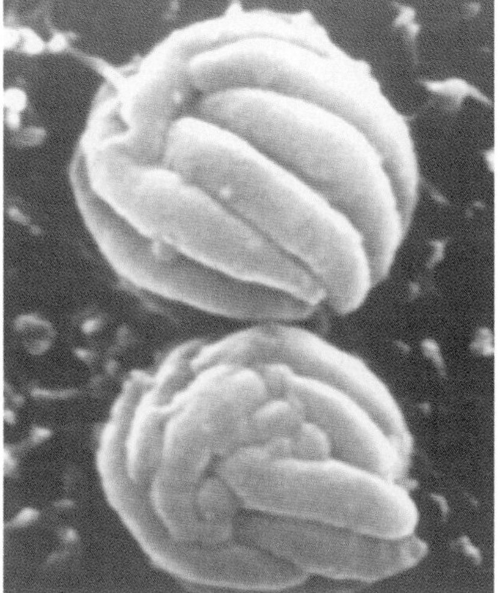

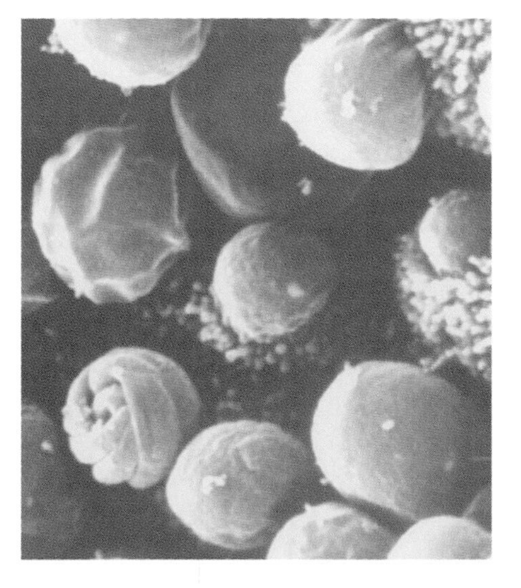

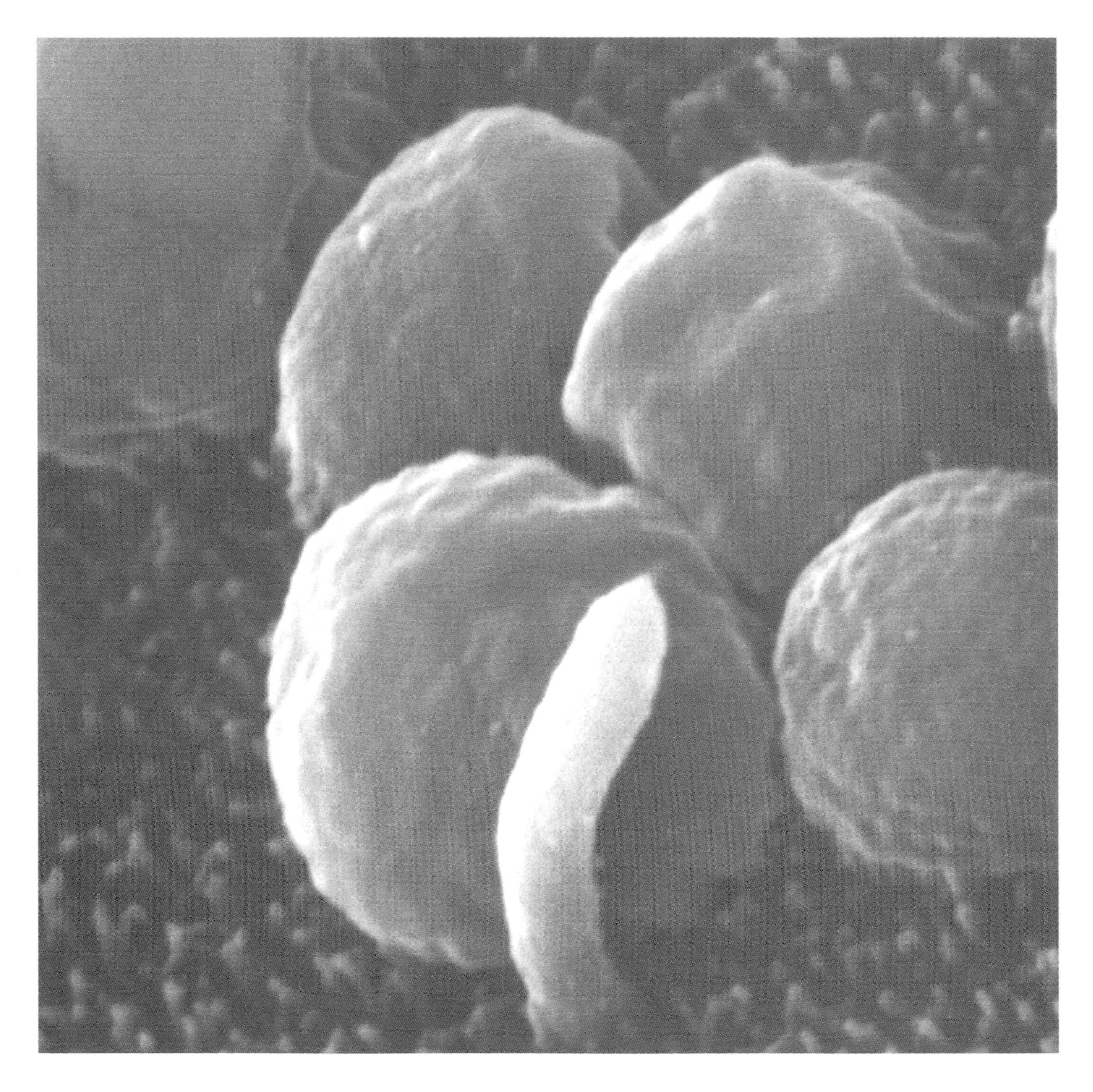

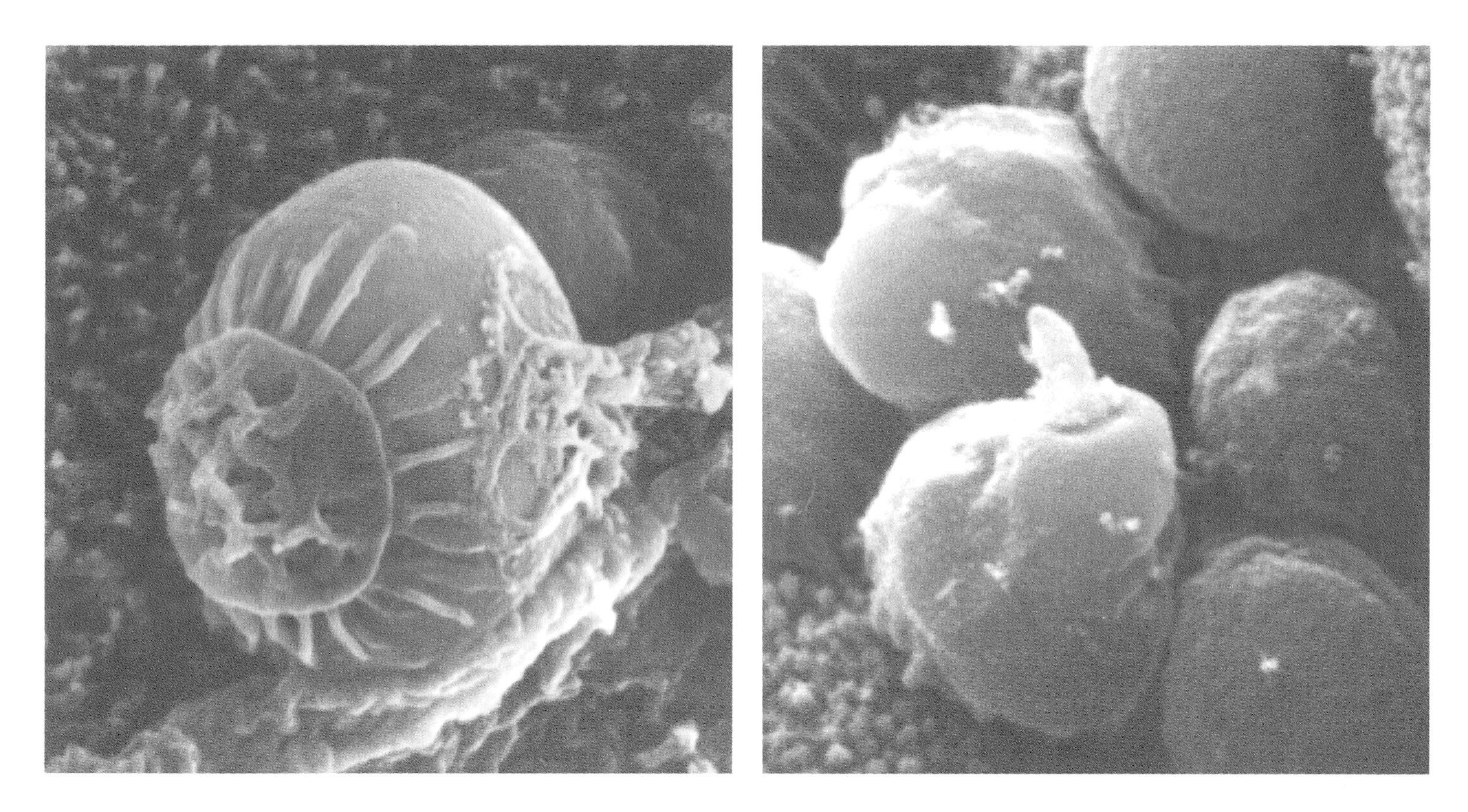

was humans or farm animals. In any event, the concentrations of the chemicals used in the water purification process were just not high enough to deal with the contamination, particularly since the contaminant—*Cryptosporidium*—is notoriously resistant to chemicals in the first place.

By the time the crisis was over, a vast number of people had been affected; the most accurate estimates put the final number at over 420,000. Worse still, there was a significant number of deaths—about 70 people died, most of whom had other medical conditions that made their health poor even before the outbreak. It was a terrible time for the city—which, in my opinion, is one of the most pleasant and pleasing waterside cities that I have visited. The modifications to the water system—including extending the water inlet by a massive 12 kilometres into the lake—have ensured that Milwaukee's water supply and active *Cryptosporidium* infections will not cross paths again.

Scary Things Nobody Wants to Know

Cryptosporidium is such a potent parasite that as few as ten organisms can give a person an attack of severe diarrhea.

Beaver Fever

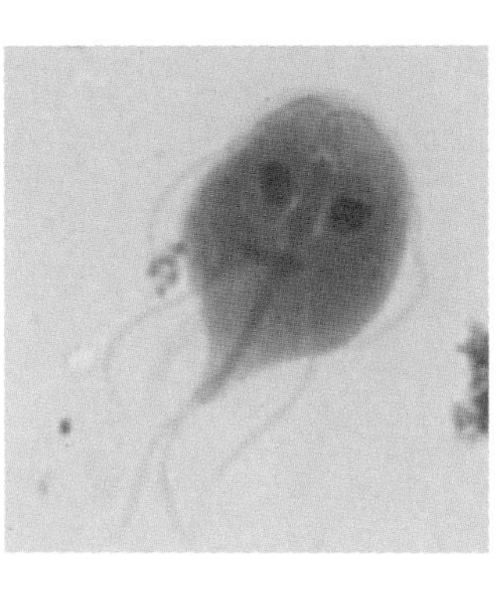

The Smile on the Face of the Tiger. It looks like a cartoon "happy face" with two big eyes and a smile, but it will wipe the smile off the face of its host. This is *Giardia lamblia,* a protozoan parasite that causes the unpleasant "beaver fever."

In the history of the fecal-oral transmission of infections, we have to be constantly reminded that the "fecal" part of the equation does not necessarily mean human feces. Animal feces can also be a problem, although we are probably less conscious of that fact in our highly urbanized lifestyles. Nevertheless, outbreaks of disease can originate from cattle feces (such as the *E. coli* 0157 infection), dogs (*Toxocara*), cats (*Toxoplasma*), and also... beavers.

The beaver is after all the national animal emblem of Canada, and I have to say that when you see them up close you can hardly stop yourself from being charmed by how cute they are. They have big teeth and a

Home Is Where the Toilet Is. This pile of sticks is what a family of beavers calls home. Unfortunately for us humans, it comes complete with an outdoor toilet—what we call "the lake."

(Opposite)
These calm and reflective waters may look pristine to the human eye, but both they and the rivers that feed it may be home to microscopic invaders.

Obvious Facts that We All Forget

Beavers are not toilet-trained.

flattish face and are incredibly industrious in terms of tree-felling and major construction. As builders, they are extremely productive. Sadly, however, the same is true of their production of feces. That wouldn't be a problem, except that beavers quite often have a parasite called *Giardia lamblia* inside them, and we can sometimes pick up that infection when we swim in a lake near a beaver dam.

Fortunately the beaver isn't a major source of *Giardia*—it can occur quite frequently in institutions such as child-care centres. The illness is usually fairly mild, consisting of a low-grade fever and diarrhea, and it often settles by itself. Nevertheless, when a microbiologist looks down a microscope and sees *Giardia* looking back, the smile is far from contagious.

Be On Your Giardia. The microvilli of the bowel are the ramparts on which the immune system attempts to keep guard—not always successfully—against marauders such as *Giardia lamblia.*

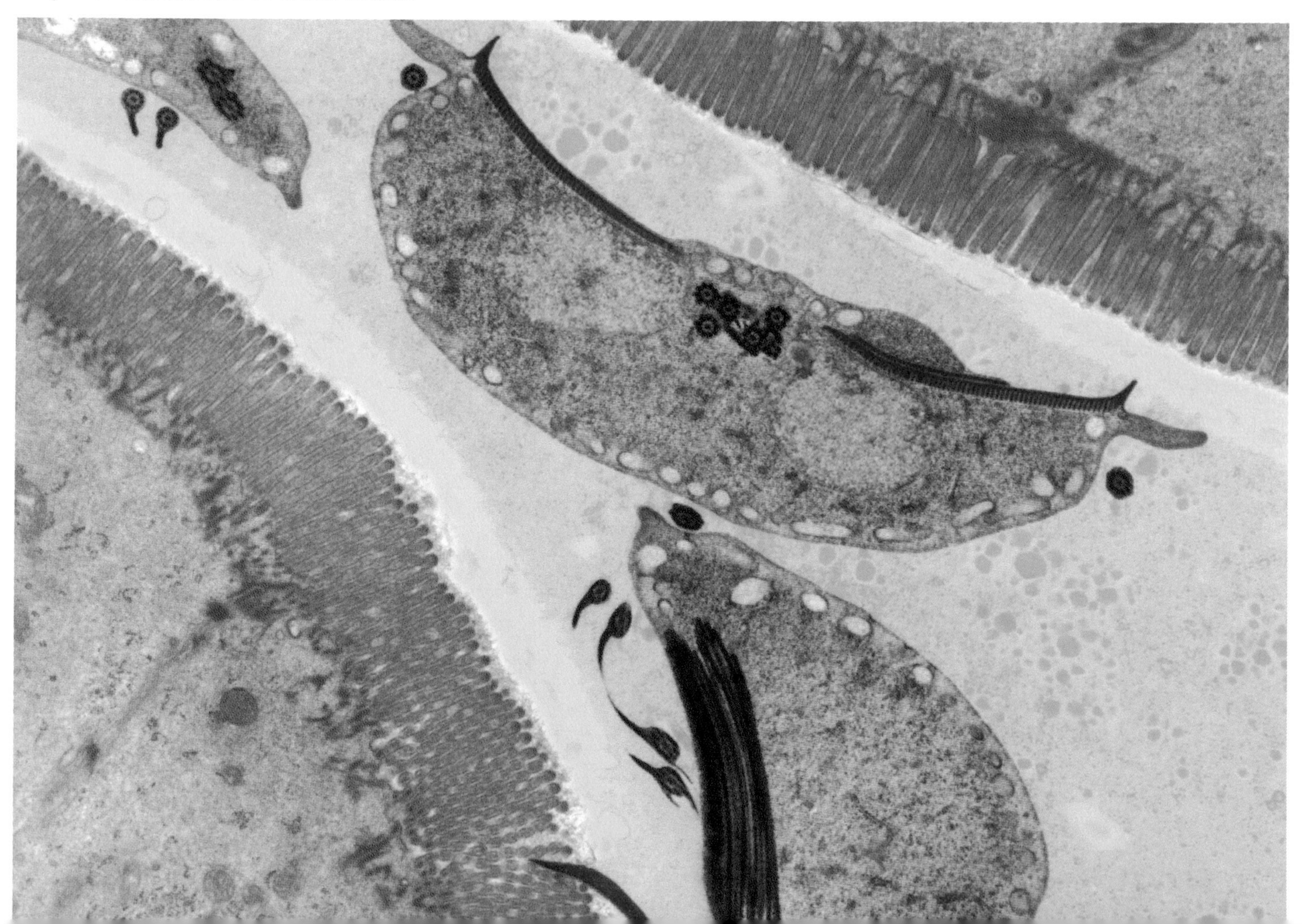

On-Line Chickens

The only time during the research for this book that I felt somewhat apprehensive and queasy was in anticipation of our visit to a chicken processing plant. Like most people who have been through medical or biological training, I do not consider myself particularly squeamish, but I did wonder about this stop on our tour. I was about to see how chickens are killed and prepared—and, by the way, if you thought that chickens are bred nowadays to grow up without heads, feet, feathers, or innards and that they crawl into plastic bags and refrigerate themselves, then I am sorry to disillusion you. (Yes, yes, I know they've now bred a chicken without feathers, but it has still got the rest of the stuff.) In the event, to my great relief, the visit was superb and I actually left feeling better—and safer—about eating chicken.

Our guide around the plant was the wonderful Tom Baker, who explained the main object of the exercise very clearly. Chickens—unlike

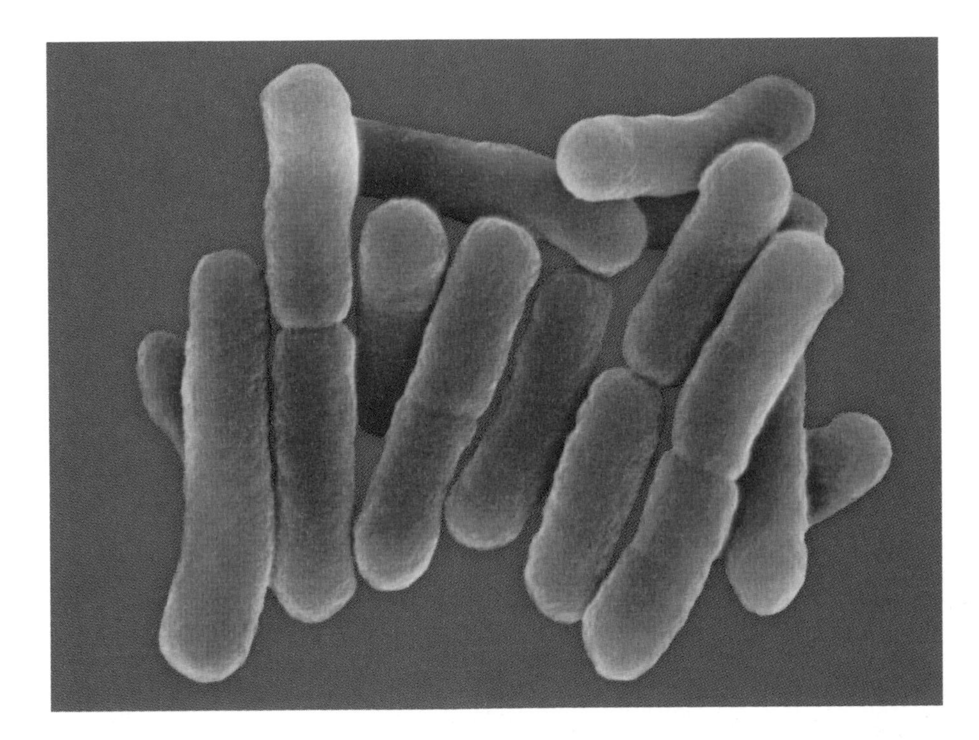

A Side Dish You Don't Want with Your Chicken. Genuinely and sincerely unwanted in human society, *Salmonella typhimurium* is a common bug in chickens.

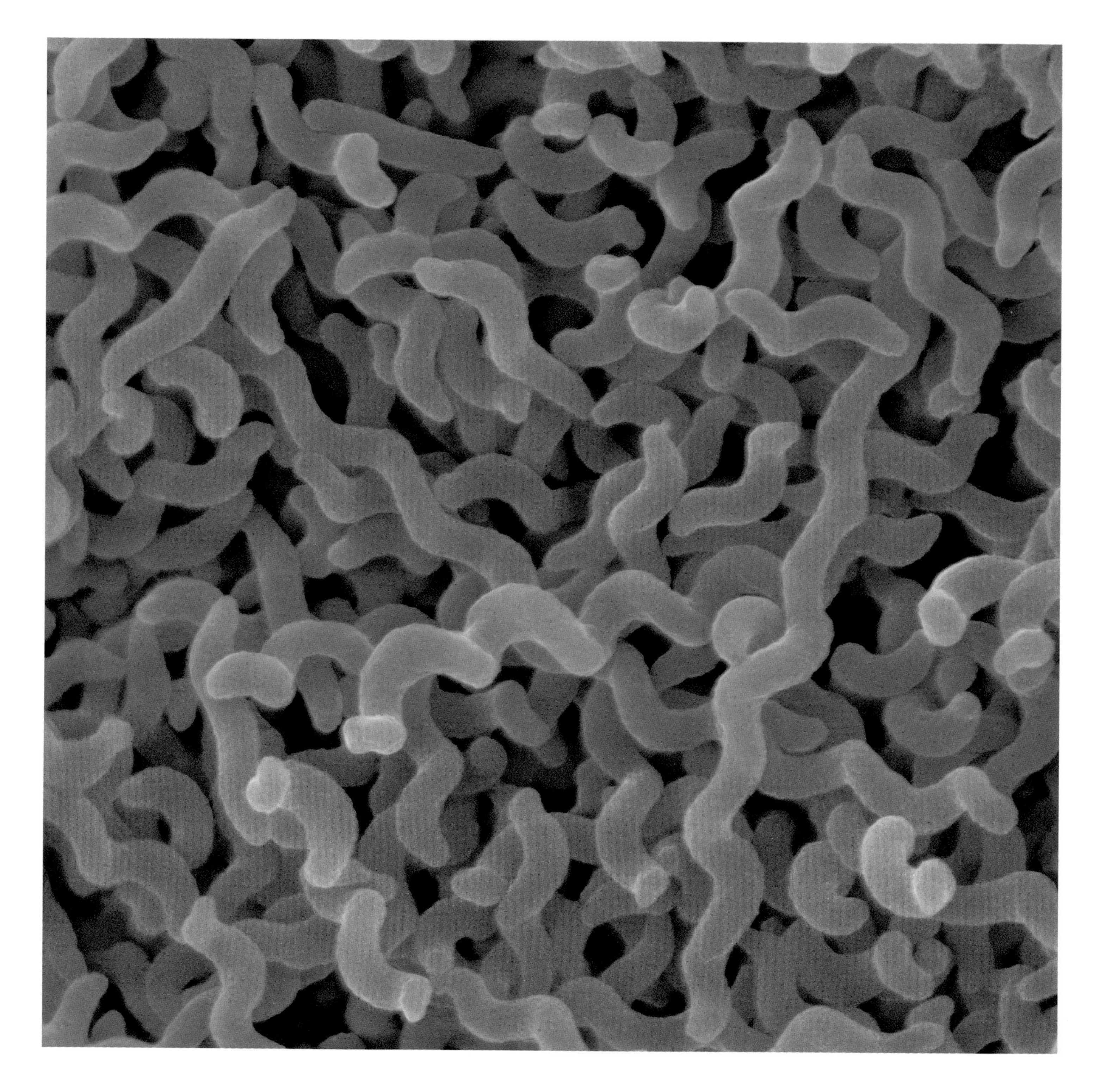

cows—quite often have certain bacteria (such as *Campylobacter* or *Salmonella*) not just on their outsides, but also living inside them. Every chicken should be thought of, as Tom Baker put it, a potential feces sandwich—potential bacterial contamination on the outside and on the inside, with the chicken's body between the two.

Let me make one thing clear: the fact that a proportion of chickens may have *Salmonella* (or *Campylobacter,* or some other bacteria) on or in them is not due to sloppy farming standards. It really is a fact of chicken life. The *Salmonella* bacteria are around—they are in the ground and they are spread relatively easily from one chicken to another wherever they are bred. We simply have to accept the fact that a certain proportion of all chickens will have *Salmonella* or *Campylobacter* inside them (which is why the careful cooking and handling of chicken meat is so important). Nonetheless, for a variety of reasons, the proportion of *Salmonella*-bearing birds is actually falling—probably about 10 per cent of chickens have *Salmonella* these days, compared to about 30 per cent a decade or so ago.

The production line for the chickens was quite civilized and not one bit as gruesome as I had imagined. Trucks arrive daily, the average truck holding about 7,500 birds in crates. They are unloaded by hand and hung by their feet on a slowly moving line of rubber-covered hooks (it is not painful). A second later the upside-down birds go through a shallow dish of water that wets the feathers on their heads and a microsecond later they receive a single electric shock that stuns them completely. From their point of view, that's the last thing they are aware of. As the line moves on, the chickens are killed and their bodies are sent first through a scald tank that loosens the feathers and then through a revolving drum

(Opposite)
Another Unwanted Condiment with Chicken. This is *Campylobacter,* a bacterium that has the unusual ability to move around instead of remaining stationary. As a result, *Campylobacter* is the commonest cause of gastroenteritis in humans worldwide.

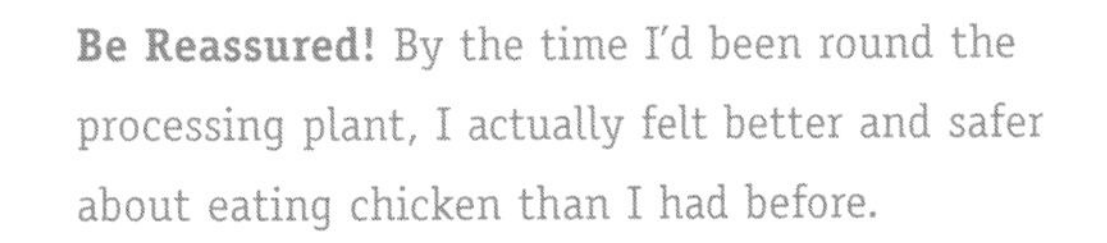

Be Reassured! By the time I'd been round the processing plant, I actually felt better and safer about eating chicken than I had before.

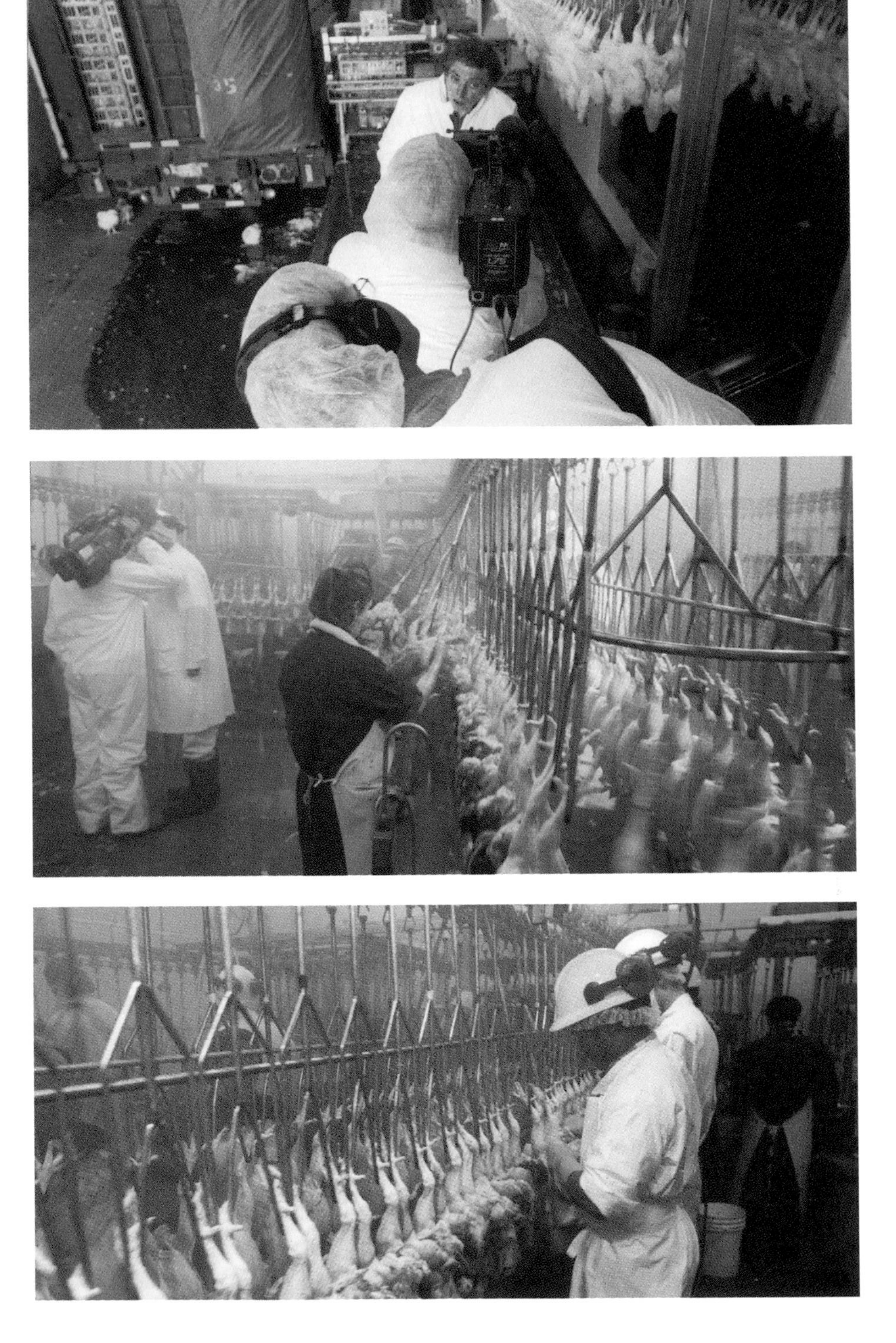

in which the feathers are separated. The carcasses are then chilled and the production line moves through a room where the heads and feet are removed, the innards are brought to the outside of the body cavity (so that they can be inspected), and the kidneys are removed from the abdomen.

This last procedure requires a machine with a lot of suction, and the only menacing feature in that whole room was the loud noise the machine made as it did its work. The carcasses are then inspected—every single carcass is inspected, both inside and out (which is why the innards have not yet been separated and are available for visible inspection). The inspectors remove any bird with any sign that raises concern, and put a metal tag on everything else. The birds are then conveyed into a chilled packing room where they are packed into plastic bags and then into large boxes.

I felt extremely reassured by what I saw there —not only because the whole process was so humane and so well thought out and clean, but also by the fact that every single carcass is physically inspected. Tom told me that some of the world's biggest plants might process more than a million birds a day—and each one can be inspected on the production line. I don't know what the opposite of the phrase "being put off" is, but if there is one, that is how I felt—I have eaten chicken several times a week since then. I make sure that I handle the meat as little as possible, that I clean the kitchen counter carefully afterward, and, if I've used the chopping boards (which I try not to do), I scrub it and dry it. I know, I know—I should have been doing that all my life, but even so, I'm glad I have now learned a little bit more about the wildlife that might be there and can take the sensible precautions.

Things I Was Always Told but Never Took Seriously Until Now

1. Have two chopping boards in your kitchen—one for meat and one for vegetables, and don't chop meat on the vegetable one.
2. After use, scrub the chopping board, rinse it with diluted bleach, and dry it.

Infections Below the Belt

Now let's stop and think for a moment. If the contents of our colons and our stool contain so many bacteria and the bacteria are potentially so dangerous—particularly if they leak and soak into our drinking water—why don't they cause problems locally, around the anal sphincter, on the skin, and, in females, in the vagina. The answer is actually very complex—and may be of very great importance in the future treatment of infections in the urinary tract and in the vagina.

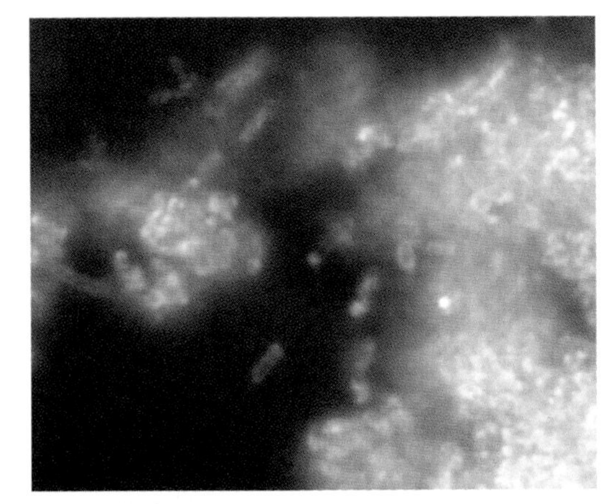

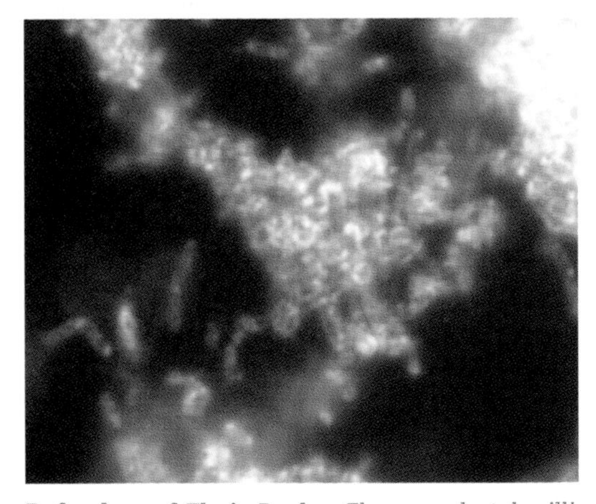

Defenders of Their Realm. These are *lactobacilli*, the "good" bacteria that keep the environment in the vagina acidic, which, in turn, discourages colonic bacteria from establishing themselves.

It has been known for several decades that, unlike most areas of the body, the vagina is an acidic environment. The lining of the vagina is, of course, well adapted to that, but it is the acidic environment that actually stops most of the itinerant bowel bacteria from establishing themselves there. It must be remembered that the vagina is anatomically quite close to the anal sphincter—and the urethra (the tube leading from the bladder to the outside world) is short and is situated in the front wall of the vagina. This means that the normally large numbers of bacteria around the anus can gain entrance quite easily to the vagina and from there to the bladder (particularly after sexual activity, for example). The "plumbing" is not well designed to prevent bacterial entry. In the male, it so happens that the anatomy is better designed from that point of view—the urethra is much longer and the penis takes the urethral opening a fair distance (relatively speaking) from the anal verge.

So it is very fortunate that the environment of the vagina is acidic and unfriendly to any wandering coliforms from the bowel. But how does it get that way? A current researcher in this area is microbiologist Dr. Gregor Reid of the Lawson Health Research Institute in London, Ontario, who is also professor of microbiology and immunology at the University of Western Ontario and director of the new Canadian Research and Development Centre for Probiotics. Dr. Reid has spent a great deal of time and research effort testing the bacterial flora of the vagina and has come up with some very intriguing suggestions for treatment and prevention of vaginal and urinary infections.

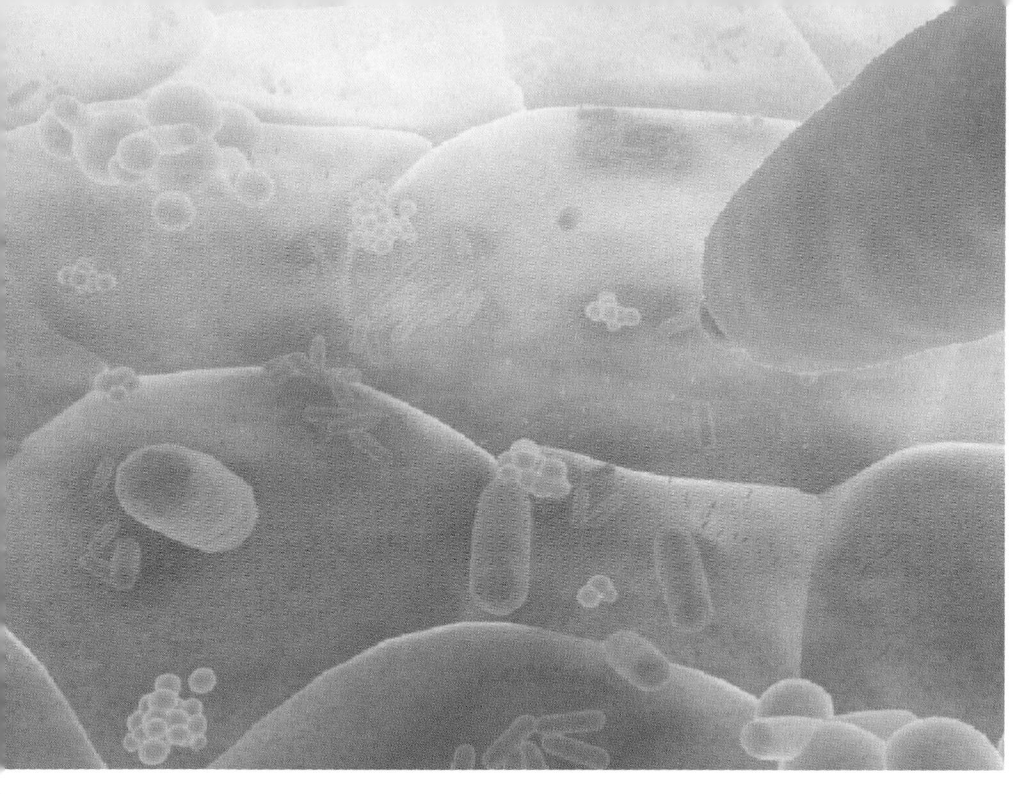

Sheep May Safely Graze. As Dr. Gregor Reid describes them, *Lactobacilli* (light blue) help to control and discourage *E coli* (red) and *enterococci* (dark blue) on the vaginal wall.

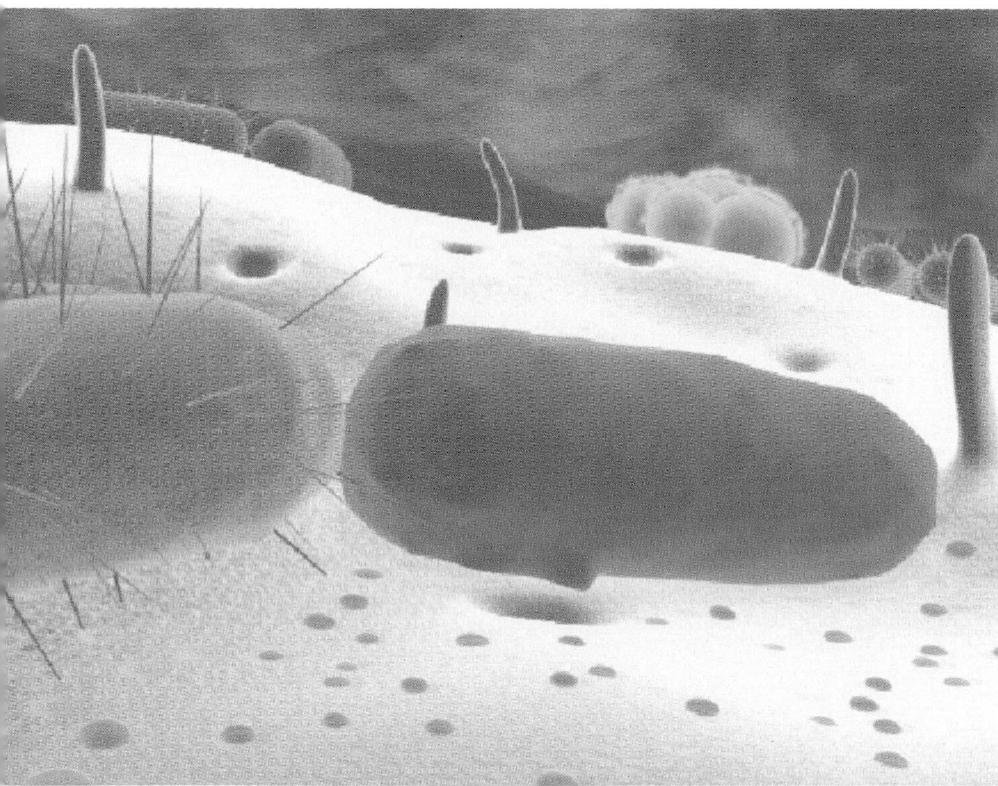

The key here are some bacteria called *Lactobacilli*. The major feature of these specimens of wildlife is that—unlike most other bacteria—their metabolism produces material of quite high acidity. Not only can they survive in fairly acid conditions, but they actually produce it themselves. In understanding why the vagina is so rarely infected despite being so close to a major source of colonic bacteria, the presence of the *Lactobacilli* is key. As Gregor Reid explains, they are like sheep grazing on

More Trouble. These are enterococci—perfectly acceptable residents of the bowel, but extremely unwelcome invaders in the urinary system or vagina.

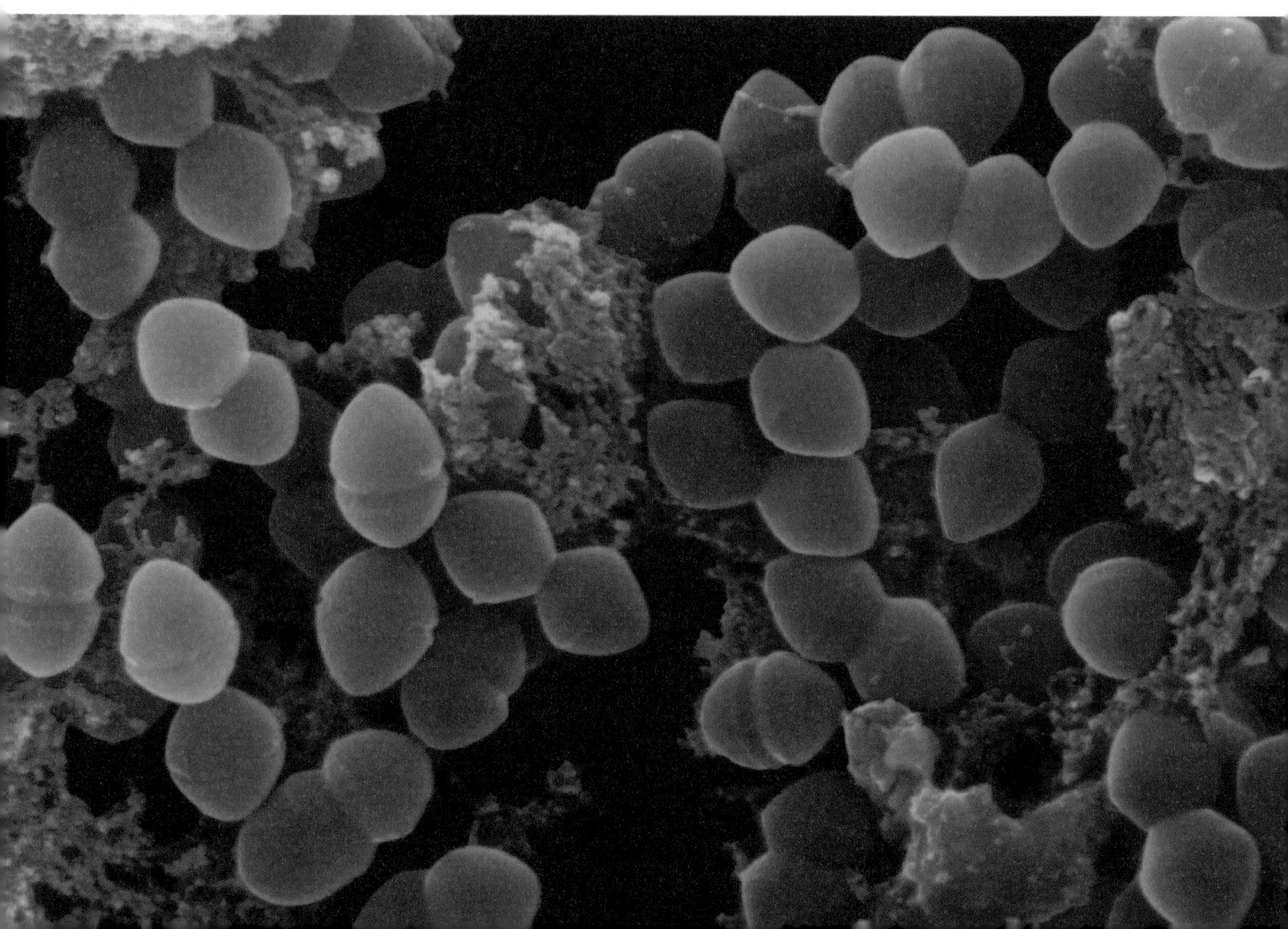

a hillside—they are both a sign and a cause of peace and tranquility. In fact, Reid's research shows that acidity is not the only defensive weapon in the *Lactobacillus*'s armoury. There are several other biological mechanisms by which they can attack, dislodge, harm, and discourage colonic bacteria—and may even help to keep other itinerants in order, such as the protozoan *Trichomonas* (which although it looks pretty when seen under the microscope is an unpleasant and nasty invader).

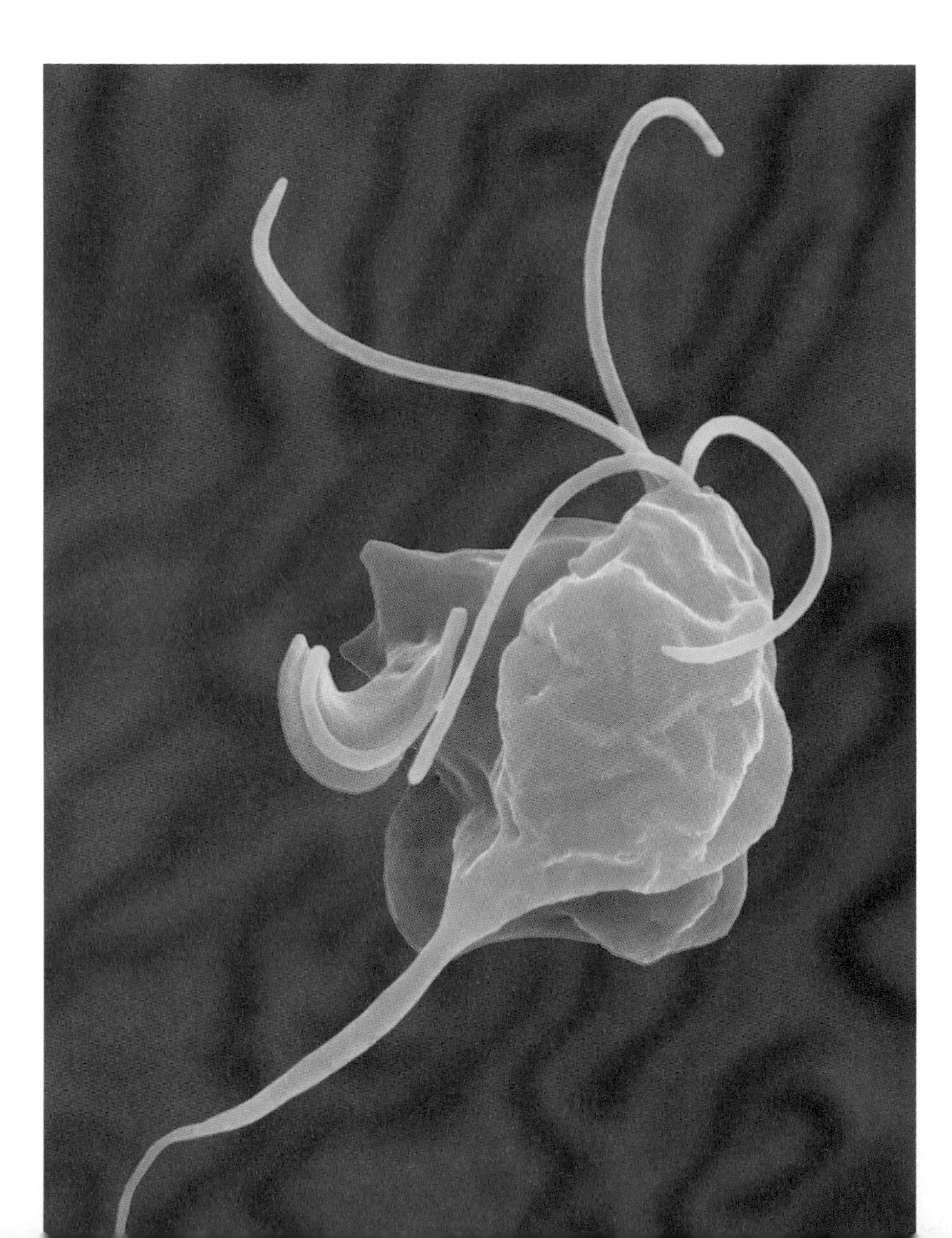

Tricky Trichomonas. This protozoan is quite a common cause of vaginitis. It is possible that acidity produced by *lactobacilli* may help to keep it under control most of the time.

Advice that Everybody Knows Anyway

Antibacterial vaginal douches are not a good idea.

All of this builds up to an area of treatment and research that is rapidly gaining momentum—the probiotic approach to treating certain infections. The principle behind probiotics is exactly the point we have been discussing here—and the one that is the central theme of this book: there is a balance between various species in an environment and it is important to know the ingredients of that balance before attempting to correct any imbalance. In the case of vaginal colonization by colonic bacteria, a more common consequence is that bowel bacteria can be propelled upwards via the short urethra into the bladder (during sexual or other activities). That means that colonic bacteria can infect the bladder—causing cystitis—or (rarely) go even higher into the kidney. The medical approach to this problem has always been to give antibiotics to kill the invading bacteria (and for serious infections—such as those of the kidney—there will probably never be an alternative). In some cases, however, not only can infections be prevented, but sometimes can actually be treated by restoring the balance, by giving the woman vaginal suppositories containing *Lactobacilli.* Systematic and large-scale clinical trials of this approach are only just beginning—but if they are successful (and my guess is that they will be), they will provide a perfect illustration of the importance of understanding the delicate balance between the co-tenants of Planet Human.

The Gifts of Motherhood

Finally, we may legitimately ask: if the bacteria in the bowel are so potentially dangerous, why do we have them in the first place? Why do we all walk around with ten feet or so of tubing inside us (our colons) full of potentially dangerous bacteria?

The answer is an extraordinary and complex one. Essentially, we have reached an extraordinary and sophisticated state of harmony with the bowel flora. These bacteria are everywhere, and we have them inside us as evidence of—and as a result of—the state of peaceful coexistence between them and us. This view is championed by many scientists in the field, including Dr. Page Caufield , whom we have already met, and Dr. Philip Tierno.

Dr. Philip Tierno, director of clinical microbiology and diagnostic immunology at the Tisch Hospital of the New York University Medical Center, is a leading expert in the relationship between humans and

Dr. Caufield

The Man Who Understands the Secret Life of Germs. Dr. Philip Tierno lives and works in New York—and remains healthy despite the multitude of bacterial threats all around him

microbiological species, and the author of a very thoughtful and thought-provoking book, *The Secret Life of Germs.* Dr. Tierno's perspectives on this subject are based on an understanding of a vast range of facts and are extremely important. As he says, germs are everywhere on this planet and we couldn't actually exist without them. It was, after all, germs—algae and bacteria—that first harnessed sunlight and made it do useful work, and accomplished other evolutionary feats such as fixing nitrogen into complex molecules. Today, they give us oxygen—although we fondly believe that we get most of this planet's oxygen from the trees in forests, the fact is that about 90 per cent of our oxygen comes from algae and bacteria in our planet's oceans.

Tierno's view—congruent with that of Page Caufield—is that humans, like most higher species, have co-evolved with bacteria. Our immune systems have evolved over the eons coping with the presence of bacteria—and the bacteria living in the bowel are an illustration of that fact, as are the bacteria in the mouth. Tierno and Caufield both point to the strong evidence that we are given our "starter kit" of bowel bacteria—as well as oral flora—from our mothers. During birth, as the baby moves out of the birth canal, there is often a small squirt of fecal material from the mother. When I was in medical school, I was always taught that this simply happened in response to the physical movement of the baby's head down the birth canal compressing the rectum as it does so. Caufield suggests that there may be more to it than that. It is possible that there is a reflex at work here—and that its function is in fact to give the newborn a tiny inoculation of normal colonic bacteria to start the colonization of the baby's gut. At this very early stage in the development of the baby's immune system, this process may actually teach the immune system to identify and tolerate "normal and natural" bacteria and to marshal its forces only when challenged by something nasty.

This isn't an easy concept to grapple with. Our instinctive embarrassment and revulsion makes us regard anything to do with feces as inherently "dirty." Yet the view that is held by those redoubtable and articulate doctors, Tierno and Caufield (among others), point to the possibility that in the right place and at the right time our colonic bacteria play a vital role in our development.

Serious but Embarrassing Hypothesis

Feces and colonic bacteria in the right place and at the right time may actually be required for the normal development of the immune system. It's only when fecal matter finds its way into the wrong and unexpected places that real problems occur.

7 Fellow Travellers

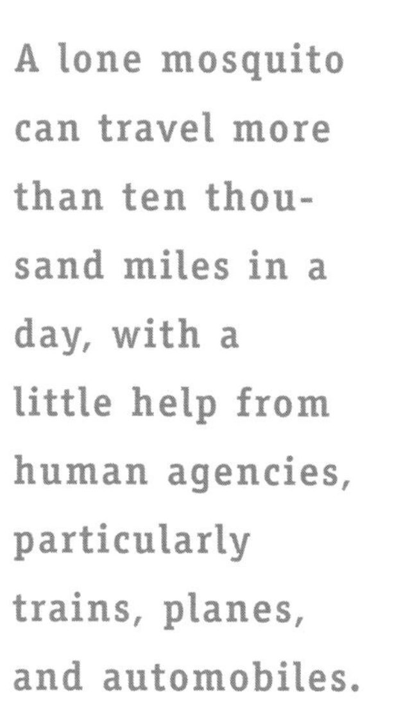

A lone mosquito can travel more than ten thousand miles in a day, with a little help from human agencies, particularly trains, planes, and automobiles.

A Leaf Taken from the Book of Life

At first I simply could not see it. All I could see was what appeared to be a thin green line starting in a hole in the ground about ten feet from the path through the tropical rain forest, and stretching out towards an old tree. As I looked up, I could see the thin green line extending up the tree until it disappeared from view in the branches about twenty feet above the ground. Then I noticed that the green line was shimmering. In fact it was twinkling. As I moved closer, I realized that it twinkled because it was moving—what I was looking at was a single file march of thousands upon thousands of little pieces of green leaf, about half an inch across, moving in a continuous procession from the top of the tree down to the ground, along a track, and into a hole. Each piece of leaf was being carried by a tiny little ant that was, at first, invisible below its load. I was looking at a highway of leafcutter ants—nature's most continuous travellers.

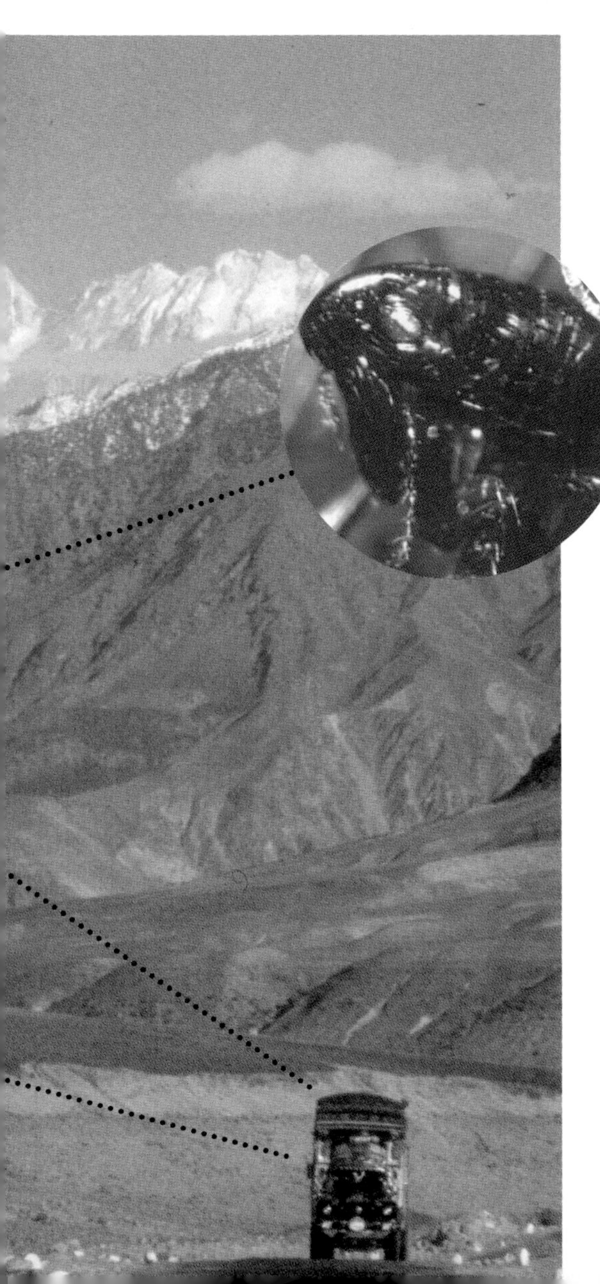

I asked our guide, Roy, what the purpose of this ceaseless and intricate activity was—fully expecting that, like most ceaseless or intricate activities in nature, it usually has to do with

Frequent Traveller Miles. Every day every leaf-cutter ant in this colony makes more than a thousand of these journeys carrying a scrap of leaf to the nest and then going back for another. Each journey is about a hundred yards and the ants' annual salary is zero. They really need a stronger union.

food or sex. In this case, it's food. The ants take the foliage down to their nest and let it rot under the influence of a type of fungus. The rotted leaf material is their food. In fact, the type of fungus is so critical to their food supply that when a queen leaves the nest to start a new one, she takes with her a few spores of the fungus. The crown jewels leave with the queen.

I noticed that on some of the leaf fragments the ants were carrying, there was another tinier ant quietly sitting on the leaf like a hitchhiker or passenger. Roy explained that the greatest enemy of the leafcutter ant is a particular type of hornet. The hitchhiker on the leaf is apparently a "spotter" ant, whose job it is to alert the column to a hornet approach. I suggested to Roy that the freeloading hitchhiker was not a spotter ant at all, but a lawyer ant. Roy thought my theory was probably wrong.

Travel Broadens the Mind and the Wildlife

The focus of this chapter is the impact of travel on the range of interactions between humans and wildlife. Fifty years ago those words would have conjured up the image of millionaires venturing to wild forests or African plains to shoot tigers, rhinoceros, elephants, or any other handsome large animal that would look good on the wall of the study. Nowadays, we realize that wildlife is a term that is not limited to large endangered species, but also applies to the entire range of organisms, many of which are small and strange, and a few of which are dangerous.

We shall therefore consider several aspects of travel. First, there is human travel, in which we venture to new places, meet new people and new species, and sometimes catch new diseases (or at least diseases that are new to us). A common example is the far-from-exotic diarrheal illness we call "turista" (or "Montezuma's revenge" or the "Katmandu quickstep" or any of fifty other euphemisms). The fact is that when we take onboard some different *E. coli* bacteria (or other species) to which the local residents are completely acclimatized, we get diarrhea. The bugs are new to us—so as the visitors, we are the ones who get the diarrhea. Travel broadens both the mind as well as our exposure to coliforms.

Sometimes—and this is just as important—we bring new diseases with us: the export of measles to several areas of the world, for example, Hawaii, has had devastating consequences.

A second aspect of the relationship between humans, wildlife, and travel is our creation of new travel destinations for the wildlife. A good example of this is our invention of air-conditioning and ventilation systems in which pools of hot water lie around ready to welcome the bacteria —a sort of spa, if you like, for the organisms that cause Legionnaire's disease.

Third, there is internal travel. Wildlife has a tendency to travel—or so it appears from our point of view. In several medical conditions, a problem in one part of the body—the stomach and the heart are good examples—has been found to be caused by a species of wildlife that we

Their Ways Are Not Our Ways. . . "Normal" *E. coli* bacteria that inhabit the bowels of humans in one area of the world do not upset their hosts, who may have co-evolved with them. Visitors and tourists who encounter those same bacteria get "turista" (aka Montezuma's revenge, the Tahiti two-step, the Delhi belly, and about four hundred other nicknames).

always thought belonged somewhere else or was just an accidental tourist. This type of internal travel or re-assignment of the role of various types of wildlife is probably going to be very significant as we research more and more deeply into many important medical conditions.

This, then is the range of travel topics with which we hope to broaden our minds. Perhaps that is one of the significant differences between the human species and the leafcutter ants—we don't stick to the path and simply trudge up and down it. We wonder and we wander. Physically and mentally we have always had an urge to explore our environment, to wander off the path. Sometimes what we find is glorious, sometimes it's dangerous—but (and I don't mean to sound preachy) it is what we make of the new information that will change things.

The Disease of "Bad Air"

As we think about travel, we need to consider the most lethal infectious disease that has ever afflicted humankind. Currently between three hundred and five hundred million people will develop this disease every year (to see those numbers in figures is even more sobering: 300,000,000 to 500,000,000). It has had many serious effects on our history, including an effect on our ability to travel and explore different areas of the world.

This parasite is so lethal that it is estimated to have killed at least half of the human beings that have ever existed on this planet—and it continues to kill literally millions of people to this day (estimates vary between 1.2 and 3 million deaths per year). Its effect on humankind is so massive that it has actually altered the genetics of the human race and has encouraged the development of a genetic disease that is itself very serious and often lethal. The disease was known in Ancient Greece, and also to the Romans, who thought—and they were almost right—that it was caused by bad air coming from the swamps; they regularly drained swamps to prevent it. Their efforts were partly successful and represented some of humankind's earliest public health measures. In Latin the words for "bad air" are *mal-aria,* which is why we call the disease *malaria.*

Even though malaria has been known for millennia, the actual cause was only discovered in the nineteenth century when in 1889 Charles Louis Alphonse Laveran identified the organism that is found in the blood of its victims (for which he received

Still Waters. Stagnant water in swamps, rice paddies, or anywhere else, can often be a haven and breeding ground for the incredibly persistent and stubborn mosquito.

the Nobel Prize for medicine in 1907). (Confusingly, the organism is not found in the bloodstream throughout the course of the disease). The species of wildlife that causes malaria is called *Plasmodium* and there are three main types of it. Still nobody was really sure how humans became infected until in 1897, in a discovery that was just as important as Laveran's, Ronald Ross identified the mosquito as the carrier—the most important one being the species *Anopheles.*

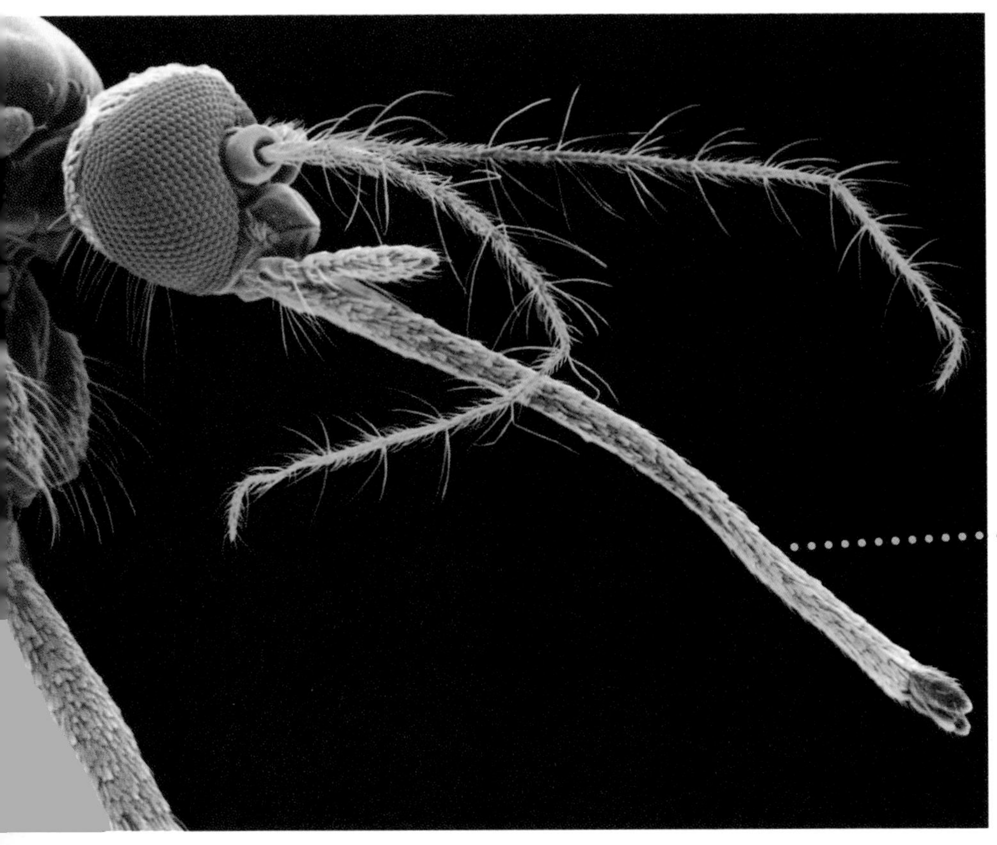

Spoozoites Enough to Make You Spit. Mosquitoes saliva may contain malarial parasites—that's not a problem for them, but it is for us.

Anopheles can carry the *Plasmodium* parasite in their stomachs, and when they bite us to suck our blood, they inadvertently inject us with some *Plasmodium* parasites, which in turn rapidly get into our red-blood cells. Inside the red blood cells, the parasites grow and develop until the red blood cell suddenly explodes with the pressure of parasites inside it. When that happens all over the body, the sudden release of a whole variety of substances causes us to have a fever attack with shivering. Worse still, the life cycle of the malarial parasite is complex and—from our point of view—very tricky in that it can hide in the liver where it becomes immune to the usual antimalarial drugs.

It is the relationship between the malarial parasite and the red blood cell that gave rise to the change in human genetics that I mentioned above. In Africa, presumably many millennia ago, a genetic fault occurred

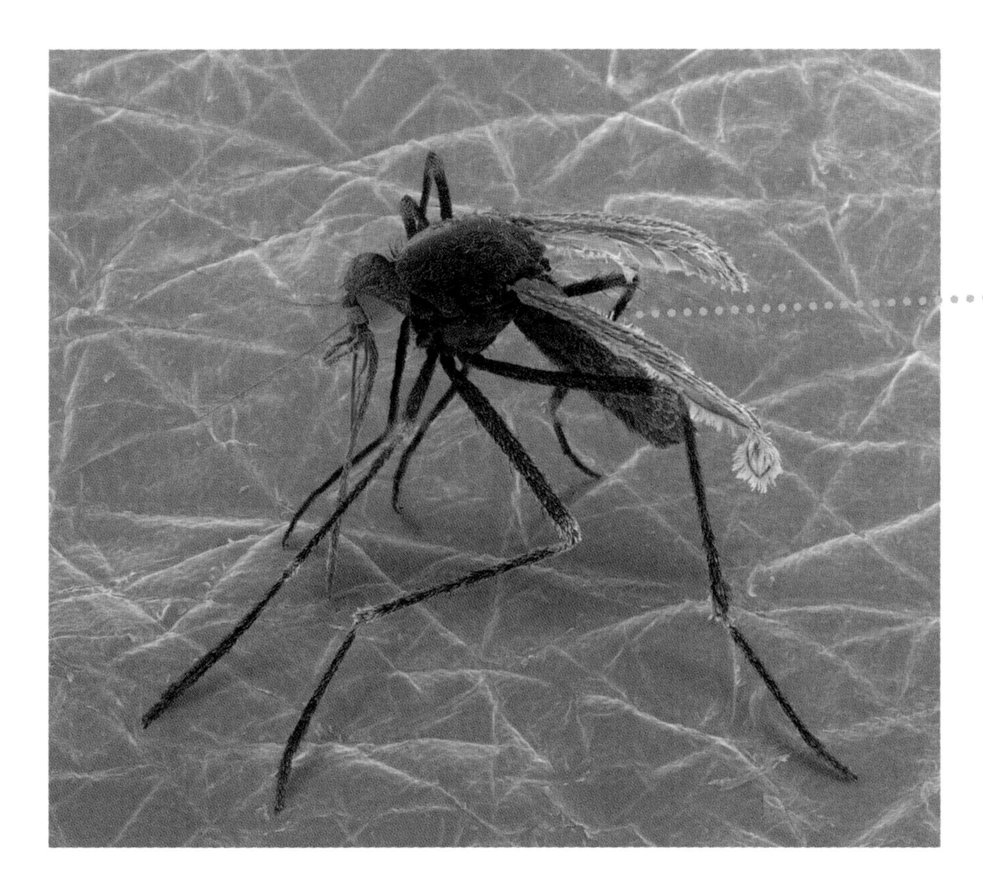

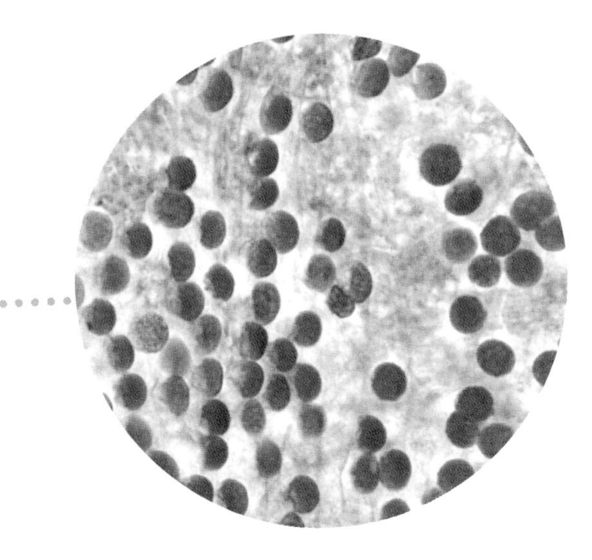

Full Service Delivery. The mosquito not only delivers the malarial parasites but also provides an environment for part of their life cycle.

by chance that produced a "mistake" in the structure of the molecule of hemoglobin—the red pigment that is responsible for carrying oxygen inside the red blood cell (and which gives it the red colour). The mistake was a tiny one—it is actually an alteration of only one single amino acid in the entire hemoglobin molecule, but it is enough to force a major change in the structure of the whole molecule. A helpful analogy might be to imagine a neat pyramid of oranges on display in a grocery store. If one of the oranges in the middle of that pyramid was replaced with, say, a melon, the entire structure of that pyramid would be drastically altered. In fact, in the case of this genetic flaw in the hemoglobin, the change in a single amino acid has such a profound effect on the arrangement of the hemoglobin molecule that it actually alters the shape of the red blood cell itself. Instead of the usual round disc, the abnormal structure of the hemoglobin twists and contorts the contour of the red cell into the shape of a crescent, or a sickle. For that reason, the disease is called *sickle-cell anemia.*

Here is how sickle-cell anemia was favoured by the widespread presence of malaria. It so happens that sickle-cells—with their disordered hemoglobin structure—cannot hold as much oxygen as normal red cells. Strangely, that confers an advantage on the person because the malarial parasite also depends on oxygen and cannot survive as well in sickle-cells. Hence people who happen to have a "single dose" of the sickle-cell-anemia gene—who have about half of their red cells twisted into sickles, and half normal red blood cells—have a survival advantage, and fare much better in malarial areas than people without the condition. This, then, explains the emergence of sickle-cell anemia in many parts of the world—it gives the people who have a single dose of the abnormal genetic material an advantage over the rest of the population.

Three Sobering Facts About Malaria

1. Malaria is the only infectious disease that is so widespread and so serious that is has affected the genetics—and the evolution—of the human species.
2. This year between 300 and 500 million people will get malaria.
3. Of all the humans who have ever lived on this planet, malaria has killed half. Currently it kills somewhere between 1.2 and 3 million people a year.

There is, however, a very serious downside. If a person with a single genetic dose of the sickle-cell gene (called a *heterozygous genetic makeup*) has children with another person who is also heterozygous for the sickle-cell gene, some of their children (one in four of them, in fact) will receive a double dose of the sickle-cell gene. They will be *homozygous* for the sickle-cell gene and that is very serious indeed. With no normal red blood cells, their health is extremely precarious; their red blood cells have a tendency to clump together and block very important arteries in many parts of the body.

From, the vantage point of the twenty-first century, it is tragic that in many areas of the developed world, there are people who just happen to

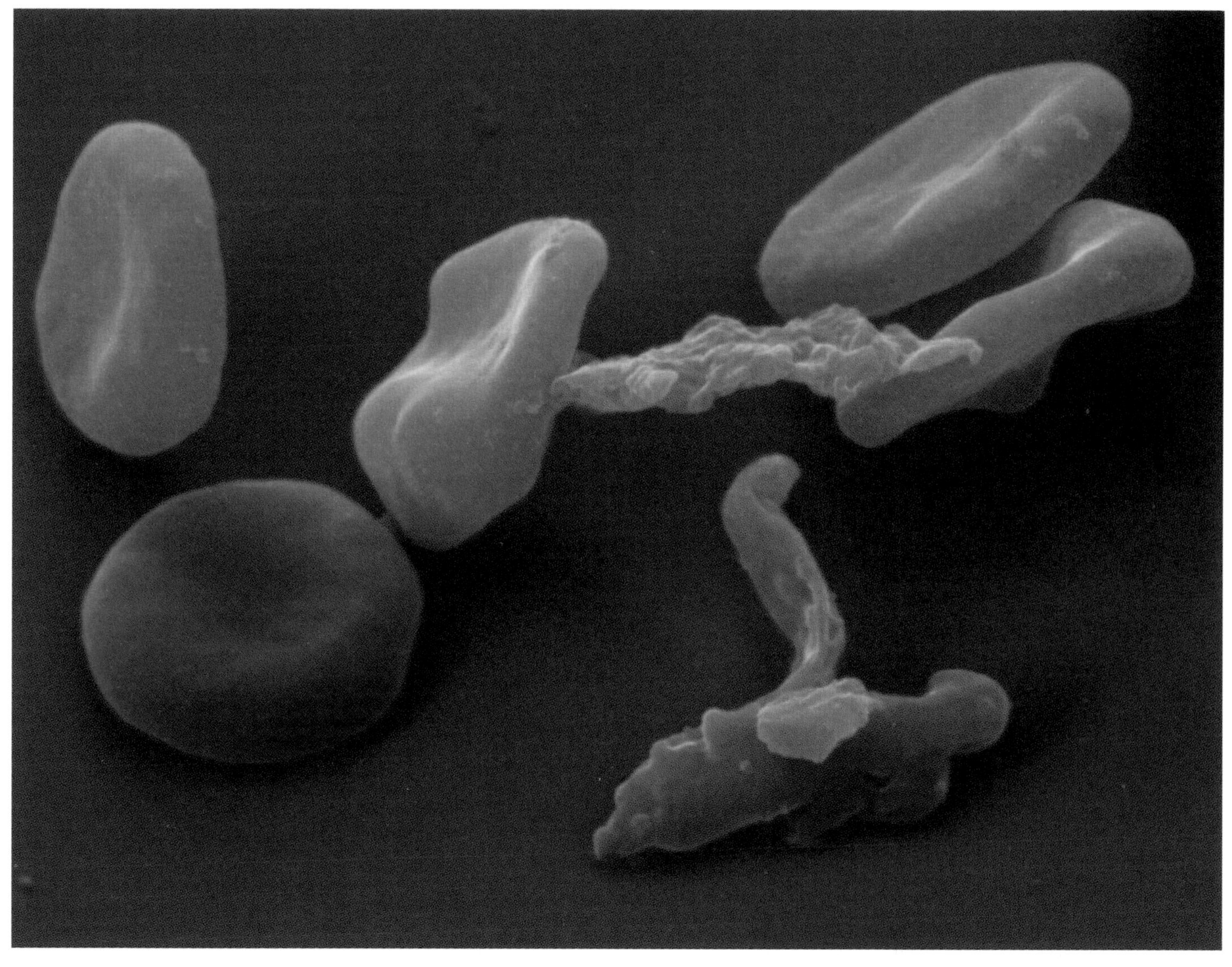

What's Bred in the Blood. Malarial parasites regard our blood cells as a limitless free lunch. Once they begin feeding and destroying cells the fevers begin.

have been born with a double dose of the sickle-cell gene and suffer incredibly (and sometimes die) as a result. It seems a genuine injustice that a genetic flaw that offers a survival advantage in malarial regions can cruelly affect innocent people on the other side of the planet. It is—in a strange and heartbreaking way—a testament to the power and continued presence of the disease of "bad air."

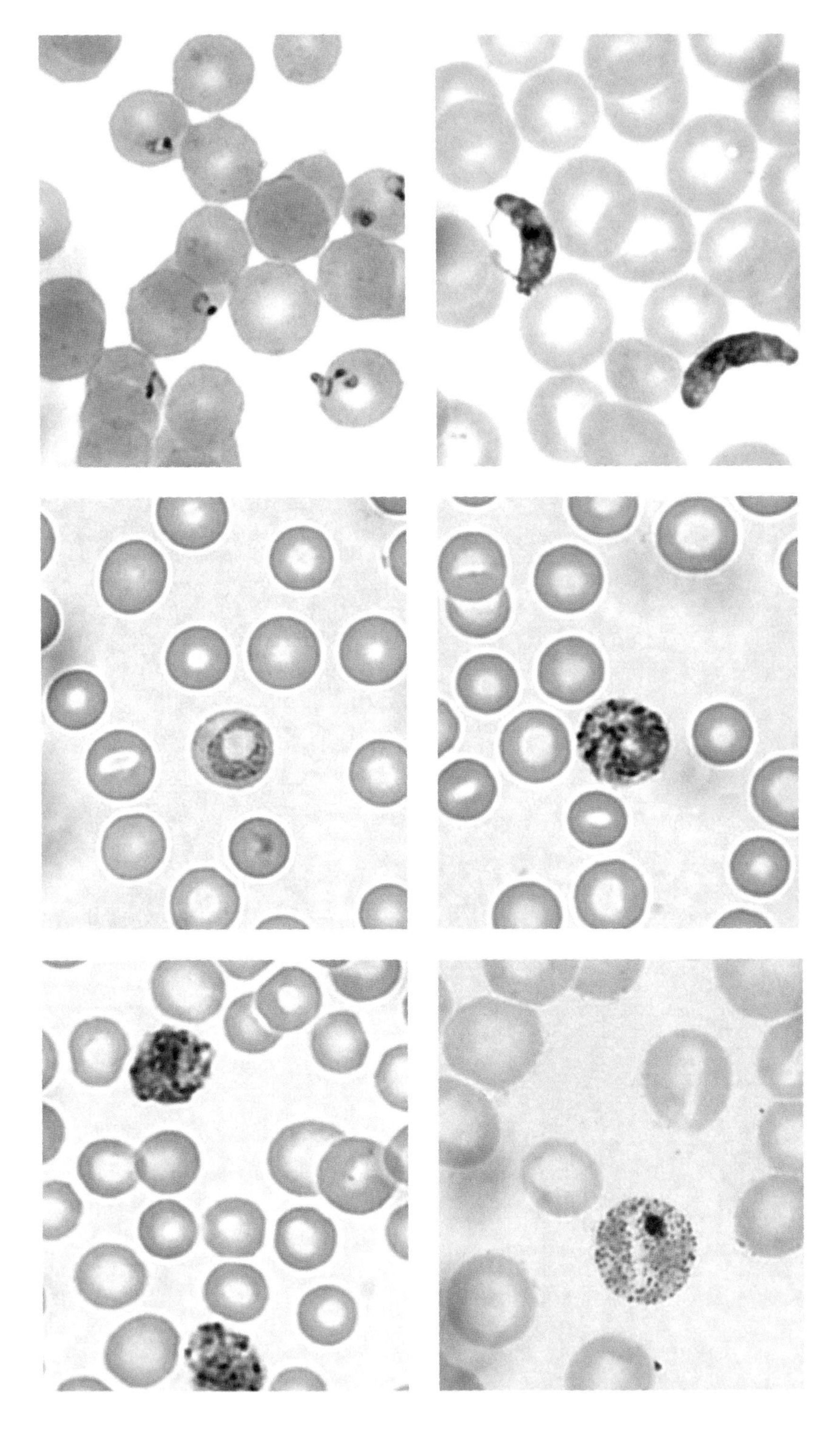

Different Faces of the Same Villain. *Plasmodium* changes its shape and location once inside a human and in doing so remains a step ahead of our immune systems. These are various stages that *plasmodium* goes through once in our blood.

Very New Strategies Against a Very Old Disease

Eradicating malaria has been an tantalizing goal—particularly for international authorities such as the World Health Organization—for over fifty years. It seemed at first that it might be possible to achieve it.

We now have drugs that can kill the malarial parasite (chloroquine, for example), as well as preventive medications that can protect people who travel into endemic areas. In retrospect, these approaches to treating the actual disease of malaria have generally been very successful—large numbers of people who would previously have died of the disease now survive. But, as the saying goes, an ounce of prevention may be worth a pound of cure. On the low-tech end of prevention strategies, it is estimated that the cost of providing every at-risk human with something as simple as a mosquito net would only cost about $5 per person per year.

A more useful strategy, you say, might be to attack the mosquito. After all, as I said at the top of this chapter, mosquitoes can travel great distances—particularly with the help of mass travel and airplanes. The problem here is that there are thousands of species of mosquito from three main genera. Of the three main genera (*Culex, Aedes,* and *Anopheles*) only species of the Anopheles mosquito are capable of transmitting malaria to humans and monkeys. There are hundreds and hundreds of species of Anopheles out there buzzing around, but out of those hundreds there are fifty to sixty species that can carry malaria parasites. If we knew where those species lived and into which areas of the world they had penetrated, we could concentrate our efforts.

This is where the high-tech world can help us. As a spin-off from military technology, there now exists what is really a portable computer that is programmed to give the wearers a picture catalogue in front of their very eyes. The man guiding this highly unusual project is Dr. Anthony Guiterez, a research chemist in the Entomological Sciences Program of the US Army Center for Health Promotion and Preventive Medicine in Bethesda, Maryland. As you can see from these pictures, soldiers have

Dr. Anthony Guiterez, US Army.

Dr. Guiterez uses a portable lab based out of the back of a Humvee to determine whether the captured mosquito carries any malaria DNA.

been equipped with a computer/receiving station that they wear on their belts. The system feeds into a special visor that has a screen providing the wearers with an image they can see right in front of them (this is called a *heads-up display*).

The beauty of this system is that it can instantly turn every wearer into a world authority on mosquitoes (or on whatever species is currently in the computer). The wearer goes out into a swamp and picks up a single mosquito. In the old days, the specimen collector would have had to write down where the insect was found, and then carry it back (with dozens of other samples, of course) and identify it. With the heads-up display, the collector can now hold the mosquito in one hand and, with the other, call up the images of various species of mosquito from the computer database, rotating the images on the visor display to get an exact match. In this way, the person using the system can instantly and precisely identify every single mosquito without even leaving the swamp.

This may turn out to be a very powerful tool indeed. With the incredibly accurate system of tracking the precise species of mosquito and its whereabouts, we may be able to learn about the migration patterns and movements of the malaria-bearing ones. This will allow us to focus on the right strategies: whether that involves draining the right swamps—as the Ancient Romans did —spraying, or some other strategy. In the ongoing struggle against malaria, as is true in so many endeavours of humankind—both explorations with peaceful intentions, and those with hostile intentions known as wars—the difference between success and failure often depends on how good your maps are.

Trypanosomes—Serious Grief

You may not have heard of the group of parasites called *Trypanosomes*—not many people outside the field of infectious diseases have—but they are an example of what protozoan (single-celled) parasites can do. You can see from the photographs on the next page that Trypanosomes look a little like the malaria parasite, but although they don't pose as vast and as lethal a threat to humans as malaria, they are very common in some areas of the world. The African species causes the serious brain disease that we call *sleeping sickness;* in South America, the organism causes a particularly bad disease called *Chaga's disease.*

Trypanosomes can live in an impressively wide range of hosts—including cats, dogs, raccoons, opossums, and armadillos. They get into humans via insects. These insect vectors are a type of tick called *reduviid bugs,* or "kissing bugs," because they tend to bite humans on the face. They are also called "assassin bugs" because of their tendency to bite other insects and kill them. The reduviid bugs hide during the day in cracks and crevices in the house, and come out at night to bite their selected human host. They excrete the Trypanosome parasites in their feces, and—unfortunately for us—the feces get into our systems through the bite that the bug gives us.

In Africa, sleeping sickness was so named because it causes marked swelling of the brain (encephalitis) and induces a form of coma or stupor in which the patient is hardly able to stay conscious. It is a very serious condition and is often fatal. The South American form is a little gentler,

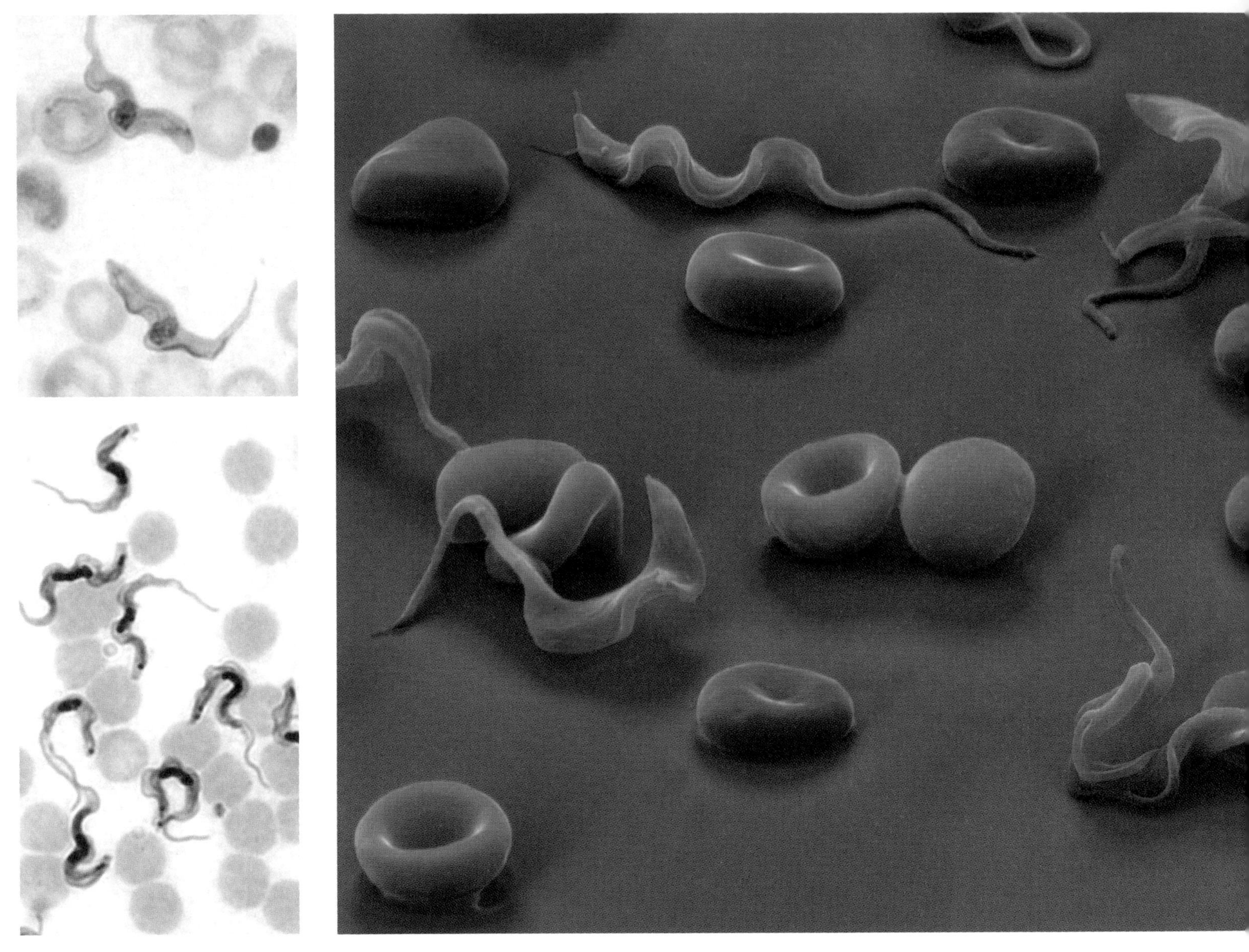

Protozoan Savages. Trypanosomes like these cause the African encephalitis called "sleeping sickness" and the South American condition known as Chaga's disease, which can affect the heart.

but not very much. The acute phase of Chaga's disease, with swollen glands and fever, from which the person usually recovers, is often followed by chronic heart problems in which the control mechanisms of the heartbeat may suddenly fail, causing death from what looks like a fatal heart attack.

New Homes for Old Wildlife

A few decades ago it would have been unimaginable that a citizen of a North American metropolis could suffer from the effects of botfly larvae. What has changed to make this not only imaginable but real is (as the theme of this chapter suggests) the accessibility of travel—we humans can now regularly cross vast distances and meet new wildlife. But a similar process is also available to the wildlife—we humans have created new worlds or environments for them to move into.

A good illustration of this problem, as I described at the beginning of the chapter, is the bacterium *Legionella*. This bacterium received very little attention until a sudden epidemic of severe pneumonia at a convention of the American Legion in Philadelphia in 1976. It was so sudden—29 people died—that, at first, it was not even certain that it was an outbreak of infectious disease; there was a serious possibility that it was in fact a case of mass poisoning. With some swift and extremely valuable research the culprit was identified as the one shown in the electron microscope picture on the next page—the bacterium *Legionella pneumophila*. It turns out that this bacterium was not new, but had actually been around for some time. What made identification difficult is that *Legionella pneumophila* is hard to see under the microscope unless a special staining technique with silver is used to render it visible.

In a way, we brought the disease now known as Legionnaire's disease upon ourselves—albeit inadvertently and with the best of motives. In the development of human civilization we have invented air conditioning and humidifiers and other methods of processing water and humidity to make our living and working environments more comfortable (and, in some cases, bearable). These systems often include places where hot water can collect and, as it happens, *Legionella* is one of the few bacteria that can survive in hot water. It thrives on it. And flourishes in the total absence of competition. Worse still, our humidification systems often turn the water into a fine mist and pump the droplets out into the area being air-conditioned. That is why we see sudden outbreaks of Legionnaire's

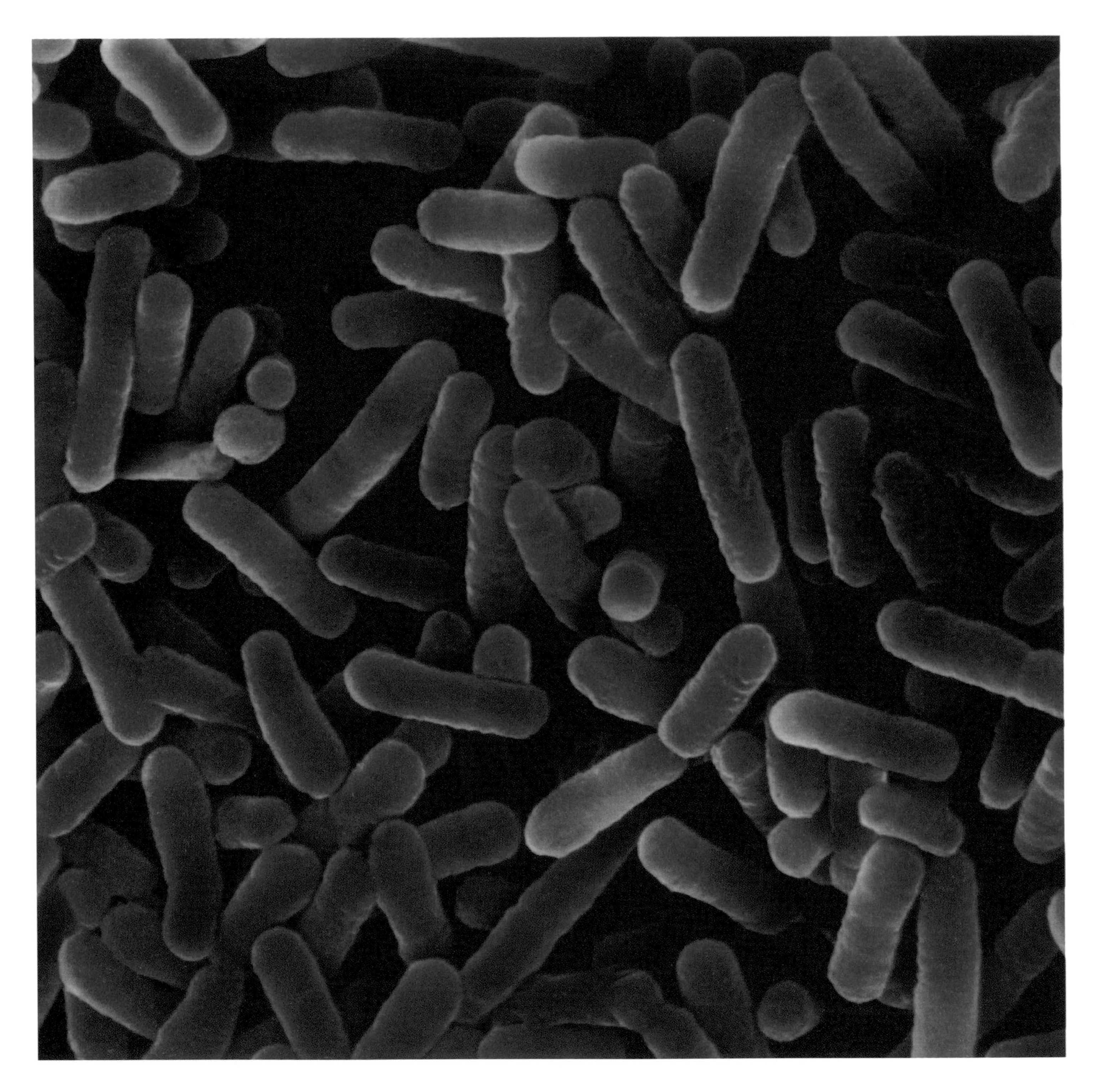

disease in a group of people who have all been together in the same air conditioned or humidified room—it has even occurred once or twice in airplanes.

Legionnaire's disease can be very serious and even lethal, particularly among the elderly. Although we all think of it as a "new" disease—as if the bacterium suddenly evolved from nowhere—it has been around for millennia. Our creation of salubrious new resorts for it in the middle of human communities is the factor that has brought these travellers into our midst.

Enough to Give You Ulcers

Everybody thought that they knew all there was to know about ulcers—they are caused by stress, by smoking, and by spicy foods. So when an Australian pathologist, Dr. Barry Marshall, a research professor of microbiology at the University of Western Australia in Perth, dared to suggest in 1983 that they were actually caused by a species of wildlife—a previously known innocuous-looking bacterium called *Helicobacter*—a lot of people thought he was stark, staring bonkers.

He wasn't; he was right. In retrospect, we (in the medical profession and in the general public) were all a bit like the pre-John Snow people who thought that cholera wasn't caused by something in the drinking water. It was just a fact of life, we thought. Just One Of Those Things. That same deep-seated feeling permeated our attitude to stomach ulcers—some people got them, some didn't, and if you smoked and led a very stressful life you'd just increase your chances of getting the damn things.

(Opposite)

Ills, Not Pills. This looks like a photograph of some capsules of a newfangled antibiotic. In fact they are the exact opposite—an extremely and insidiously unpleasant species of bacterial wildlife called *Legionella*.

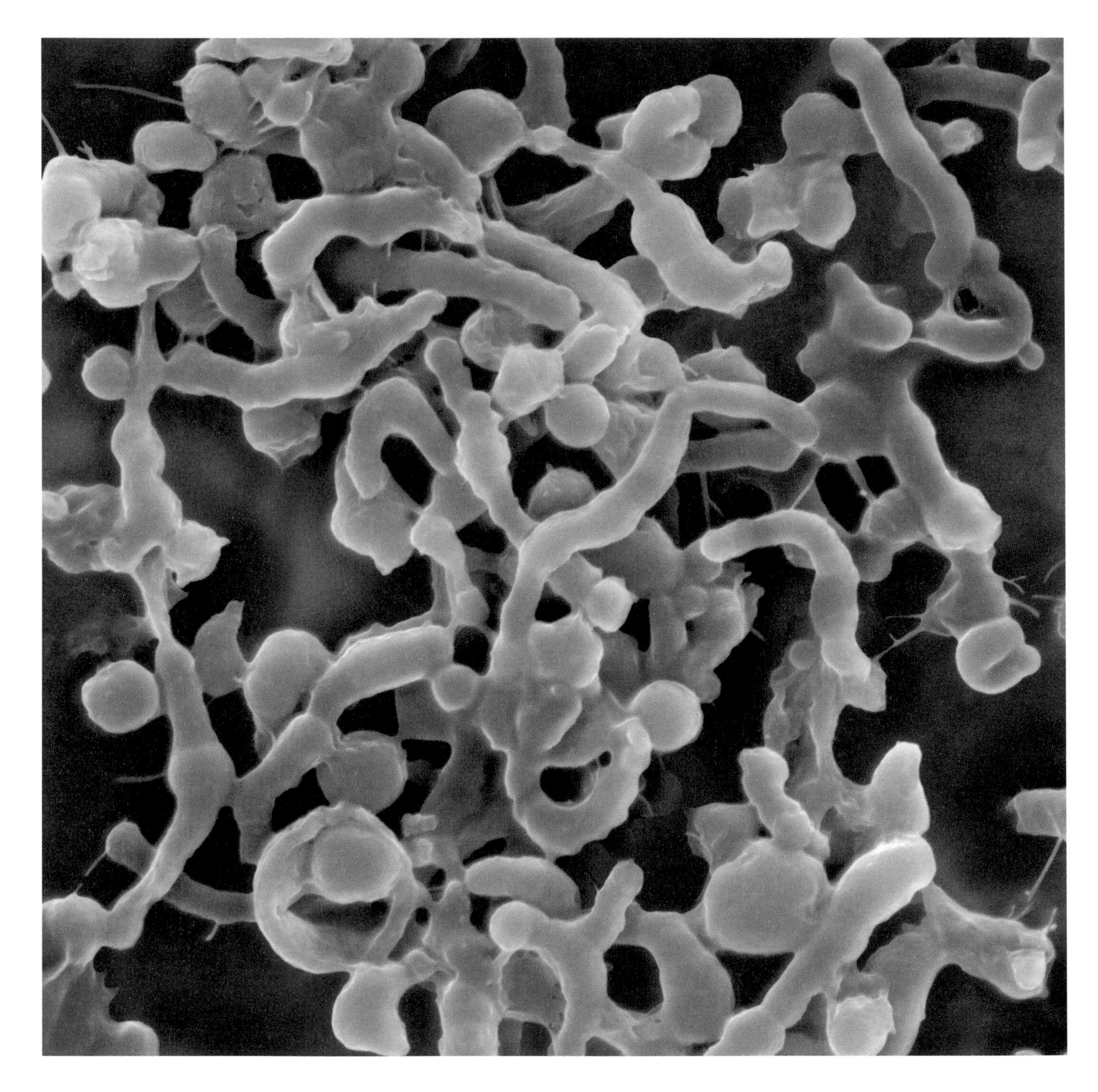

I discussed the sea change in our understanding of stomach ulcers with a friend and colleague of mine. Dr. Laurence Cohen at the Sunnybrook and Women's Health Sciences Centre in Toronto. For a physician in the front line of ulcer treatment, Laurence certainly has a very clear perspective on the revolution in thinking Barry Marshall brought about. When Marshall started out, he was perceived by almost everyone as a heretic—a man outside the mainstream of clinical medicine—a pathologist (not even a genuine and proper patient-treating doctor!). But even so, Marshall started noticing that present in all the biopsies of stomach ulcers he looked at was a rather puny-looking *bacterium—Helicobacter.* As far as the rest of medicine was concerned, *Helicobacter* was just another casual traveller passing through—it had nothing to do with the stomach and nothing to with the ulcer.

Marshall wasn't the first person to notice the presence of these seemingly insignificant bacteria, but he was the first to take real notice of them. Everyone had always assumed—as I was taught at medical school—that *Helicobacter* was a pretty pathetic sort of creature that just happened to have the unusual property of being relatively resistant to stomach acid. It was no surprise (and no big deal)—the reasoning went—that *Helicobacter* was present in the stomach because it was obviously the only traveller left standing after all the other bugs that we swallow in our food had been killed off in the acid bath. To blame the ulcer on *Helicobacter*—as Marshall seemed to be implying—was as daft as blaming firefighters for starting fires (*Every time there's a fire, those firefighters are there—it must be them that's doing it!*).

Despite the increasingly frenetic flak, Dr. Marshall steadily accumulated his database of evidence and then did something that only a very few physicians have done before him. He gave himself a stomach ulcer by swallowing some *Helicobacter* bacteria. To ensure that his experiment was watertight, he arranged to have a gastroscopy to prove that he had no previous stomach ulcers and had some specimens taken from his stomach to prove that he had no *Helicobacter* either. Then he deliberately gave

(Opposite)
Culprit, Not a Casual Visitor. This is *Helicobacter,* a bacterium that happens to be able to live in an acid environment such as the inside of the stomach. It had often been seen in stomach ulcers, but was never thought to be the cause of them.

himself a dose of the *Helicobacter,* and, after an appropriate amount of time had elapsed, arranged to have a second gastroscopy. This time the tests showed clearly that he had a stomach ulcer and there were *Helicobacter* organisms present. To clinch the case, he took a course of antibiotics that would kill the *Helicobacter* and organized a third gastroscopy. The ulcer had healed and the *Helicobacter* were gone.

According to the Laws of Disease Causation set down by Robert Koch (including, among other criteria, that the organism must be seen in every case of the disease and that treatment of the organism must abolish the disease), Barry Marshall had proven his case. Furthermore he joined the very small group of physicians brave enough to try their ideas out on themselves (I believe that company includes Dr. John Hunter, who gave himself a dose of syphilis, and Dr. Seldinger, who performed the first cardiac catheterization by inserting a tube up his arm vein into his heart). Dr. Marshall proved that the tiny body-travelling *Helicobacter* organism (now renamed *Helicobacter pylorii*), when located in its new-found-land of the stomach, was the culprit for something we had been studying for more than a century.

In retrospect, we were all wrong, and Barry Marshall was a Galileo telling the world something that seemed to contradict every bit of common sense. But as the old saying goes, "there are none so blind as those who will not see." Everything in the treatment of ulcers has changed as a result of Barry Marshall's work—the special operations to cut the nerves that regulate acid secretion, the "white food and milk" diets, and the "rest-cures." These days, a stomach ulcer is predominately seen as something that happens when a bit of wildlife settles in the stomach and makes a hole. The remedy is to kill that bit of wildlife. That is good treatment for the condition—and a salutary lesson for all of us in clinical medicine.

Chlamydia—At the Heart of the Matter?

What happened with stomach ulcers and the realization that the once-innocent *Helicobacter* was now guilty beyond a reasonable doubt may or may not currently be happening in heart disease with another species of wildlife—*Chlamydia pneumoniae.* Now the Chlamydia family of organisms are rather peculiar to begin with—they are bacteria that can only stay alive if they are inside a host cell. Thirty or more years ago, the only Chlamydia that most people had heard of was *Chlamydia trachomatis*—a major cause of blindness in Africa, among other places. Then, in the late 1970s, some infectious disease experts began to link that Chlamydia with a common sexually transmitted disease known—at that time—as *Non-Specific Urethritis (NSU),* now called *Non-Gonococcal Urethritis (NGU).* This was a great surprise to the medical community, but further investigation proved that the researchers were correct.

For a couple of decades or more, we have also known about troubles caused by another member of the Chlamydia family—*Chlamydia pneumoniae.* This one causes outbreaks of pneumonia (hence its name) of varying severity. However, very recently, several researchers have identified it as a factor in the causation of heart disease—the very common kind in which the coronary arteries become furred up with atherosclerosis and the person experiences angina or suffers a heart attack or may have a combination of the two.

At this moment, we may—or may not—be in the middle of a paradigm shift similar to the *Helicobacter*-and-ulcers story, perhaps even faintly reminiscent of the Galileo-and-the-sun and the cholera-and-John-Snow revolutions. We are at least prepared this time to believe that, in the same way that *Helicobacter* is related to gastritis and the development of stomach ulcers, it is possible that a species of Chlamydia is related in some way to the beginnings of commonest type of heart disease. Of course—as we did with ulcers—we think we know all about heart disease. It's caused by being overweight, by being stressed, by smoking, by cholesterol, by selecting your parents with insufficient care (as the saying goes), and a number of other easily identifiable factors. (These observations are, by the way, still important—non-smokers have much less heart disease than smokers, for example.) But we have always known that there was something important that we didn't know. In the causation of heart disease, the known factors explain a lot—but not everything. Now it appears that one of the missing pieces just might be a species of wildlife—Chlamydia.

I discussed the very new—and quite complicated—area of current research with Dr. Bill Fong in the Department of Infectious Diseases at St. Michael's Hospital in Toronto. This is a new area of research, and the findings are still quite young, but many researchers have been able to prove that the DNA of *Chlamydia pneumoniae* can be found within the plaques of atherosclerosis—the splodges of fatty material that clog coronary arteries. In other words, the DNA "skeletons" of the bodies of Chlamydia have been identified within the tombs of the fatty atherosclerotic plaques.

At this stage, all one can say for certain is that Chlamydia and atherosclerosis are associated—the exact mechanism for that association has not been fully established. One possibility—and it's quite likely and very exciting—is that when a person has a mild chest infection caused by Chlamydia, somehow the bacterium ends up in the coronary arteries and, by some means, precipitates a deposit of fatty atherosclerosis on top of it. Perhaps it is like the grain of sand that gets under the oyster's shell and irritates the oyster into making a pearl cocoon for it. Perhaps—and this is definitely conjecture—the process of making an atherosclerotic cholesterol cocoon (or, rather, sarcophagus) for a dead Chlamydia is much more likely to happen in a person who has high blood cholesterol levels, is a smoker, is overweight, and so on. For now, we can only conjecture; the mechanisms—and any causal relationships—are still unknown.

There are, however, clinical applications that can be investigated and tested even before the exact mechanisms have been clearly established—and those studies are already under way. Studies in heart disease are being done that are similar to those that used antibiotics to treat people with stomach ulcers for *Helicobacter*. People who have already had a heart attack have been given antibiotics that will attack Chlamydia to see whether killing the Chlamydia will prevent a second heart attack. Some of these studies—but not all—are showing promising results. We do have to bear in mind that even if patients do not show improvement after receiving the antibiotics, it is still possible that Chlamydia is the cause; the key might be that you have to kill the Chlamydia before the plaques are well-established, not after.

We will not know for another few years whether Chlamydia is the missing factor in the development of heart disease, but these are nevertheless fascinating times for the researchers into the wiles of human wildlife. The role of germs in stomach ulcers is proven; in heart disease it is still pending—and who knows what next? There are so many human ailments where we simply do not know what triggers the pathological process— and clearly some of those (perhaps many) will eventually be shown to involve the activity of some form of human wildlife or another. The paths of human beings and of the other species on this planet clearly cross and intertwine in far more ways than we had ever suspected.

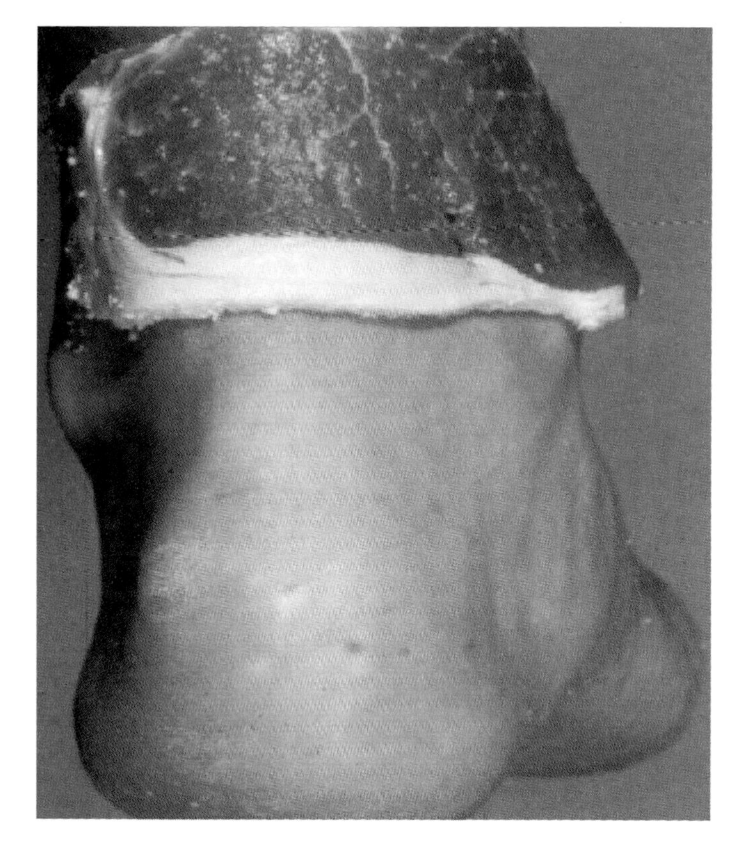

Achin' Bacon. The wildlife under the skin of this person's ankle can be suffocated and drawn out by any thick occlusive dressing. If nothing else is handy, bacon will do.

Unwanted Souvenirs of Your Travels

In the end when the travels are over, all you have left are your souvenirs. After a really good holiday you usually bring back a bunch of souvenirs that you bought at a gift shop or local market. They are often indigenous in some way, they may be expensive or cheap, they may be handmade or manufactured, they may be pretty or ugly, but they are not usually excruciatingly painful or temporarily disabling.

In the particular instance illustrated here, the souvenir—brought back quite unintentionally—was genuinely painful, and the tourist and temporary owner of the souvenir was extremely glad when she parted company with it.

The patient in this case had just returned from a holiday in Peru. She had had a great time there, but over the last few days she had noticed a swelling just above her ankle. Wisely she went to the Traveller's Clinic run by Dr. Kevin Kain at the University Health Network in Toronto. He knew the diagnosis instantly—something under the skin of the ankle was alive and it was moving about.

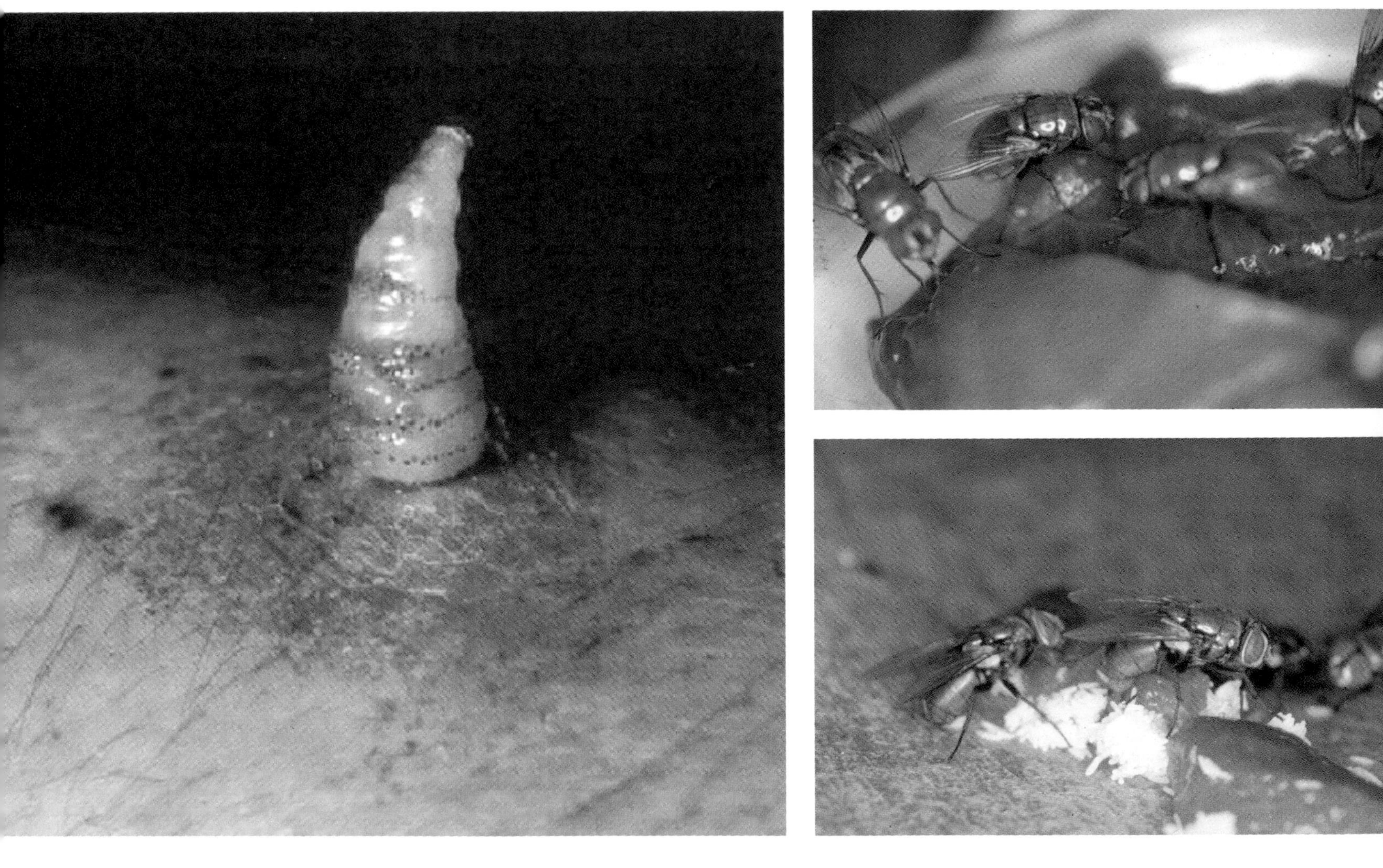

Peek-a-Boo! Lured out of the skin by the bacon wrap, this botfly is easy pickings for the attending physician.

Dr. Kain called our camera team and, with the patient's willing permission, we were there to photograph the treatment of the subcutaneous wriggling thing. If you happen to be squeamish, may I encourage you to turn the page right now and I'll meet you there?

If you're still here, then I shall assume it is because you want to be.

The thing crawling around under the lady's skin was the larva of a fly called the *botfly*. In fact, Dr. Kain suspected that there was more than one of the creatures there, and set about getting them out.

The pictures speak for themselves. There were eleven of the little monsters there (that is a record, by the way—Dr. Kain had never heard of that many botfly larvae in a single area.) The woman suffered very little discomfort—she actually got happier and happier as she and the botfly larvae were separated from each other.

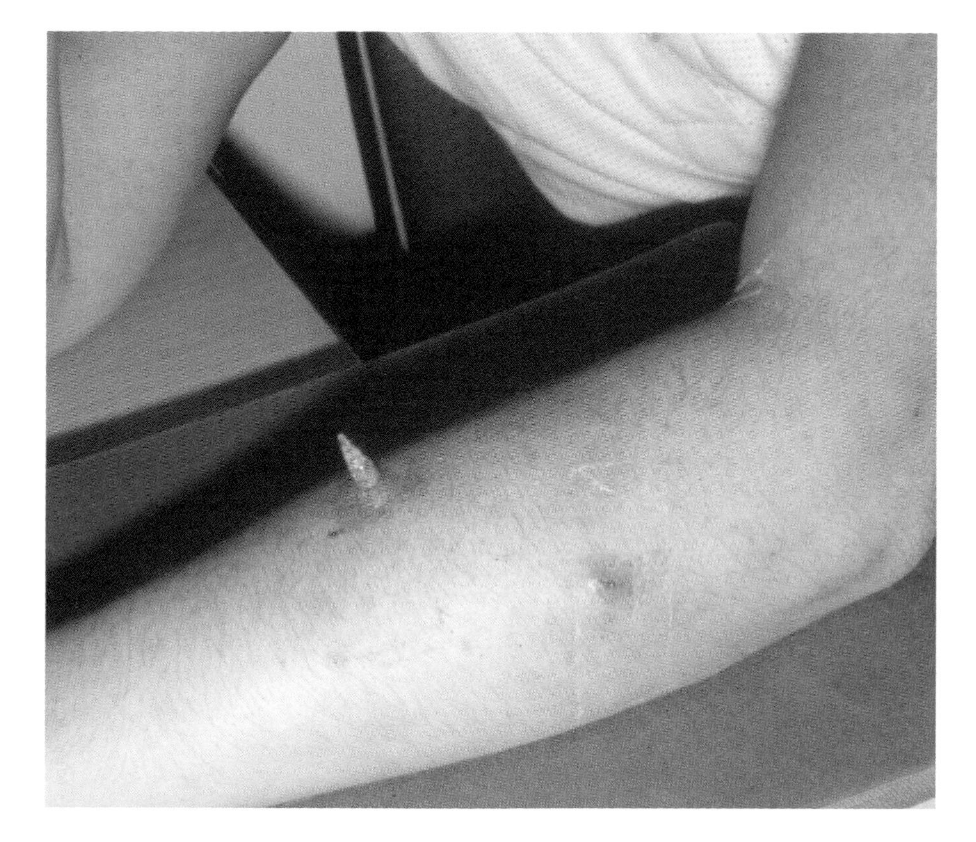

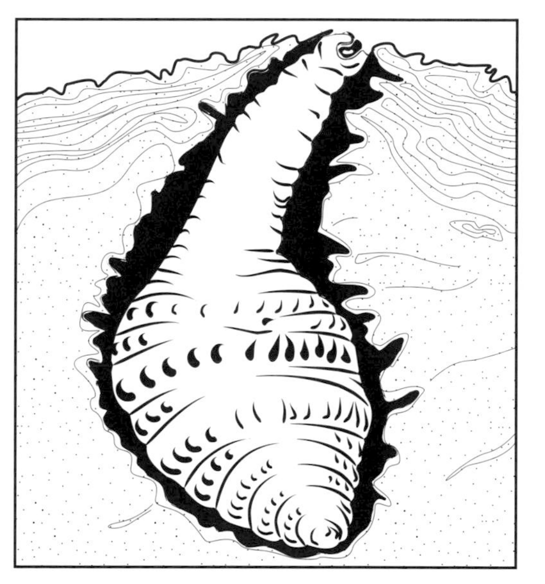

The botfly at home, in your skin.

Dr. Keystone removes the dressing with the Vaseline that hopefully smothered what he believed to be botflies.

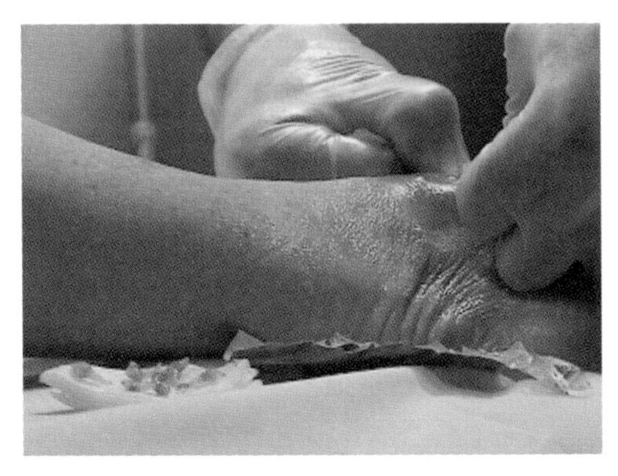

Applying lateral pressure to the bite wounds, Dr. Keystone began the eviction.

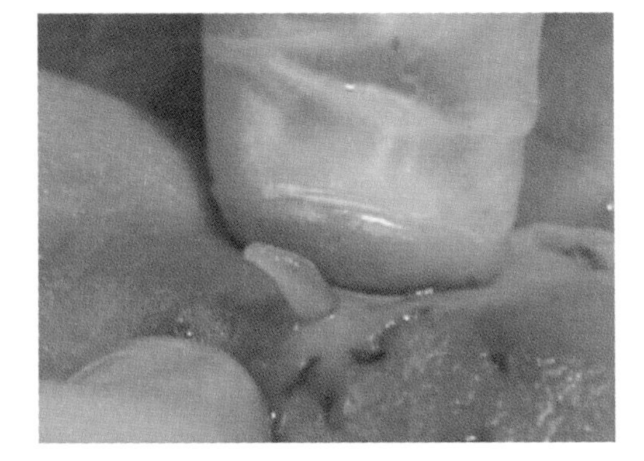

He was right! It was botfly, but the Vaseline dressing hadn't killed the invaders. But it had weakened them enough that they relaxed their grip inside the wound, allowing the good doctor to squeeze them out one. . .

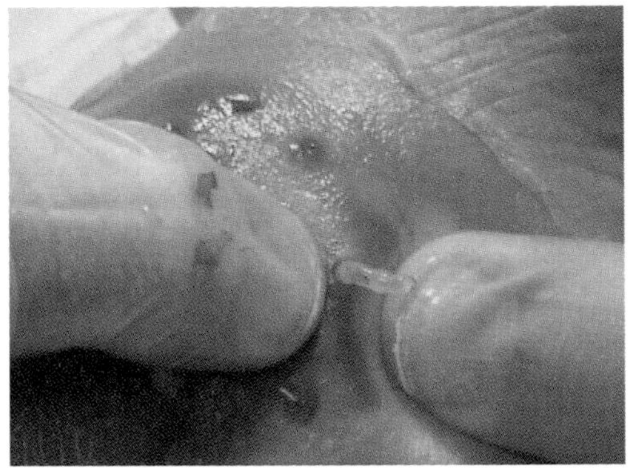

by one. . .

by one.

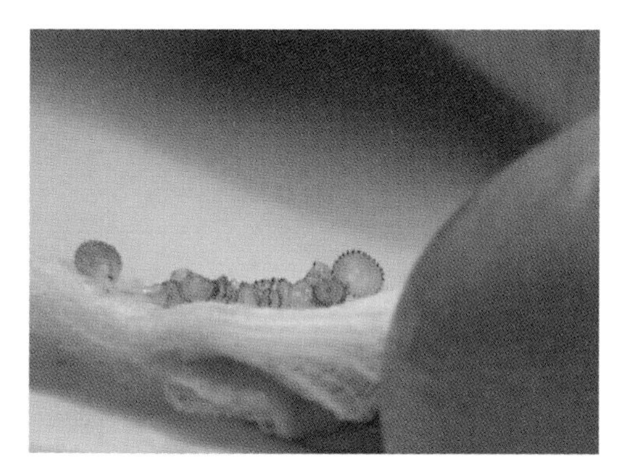

Until quite a collection were removed. Those tiny black spines are what they use to hold themselves in the wound. Once weakened, they popped out quickly enough with a little help.

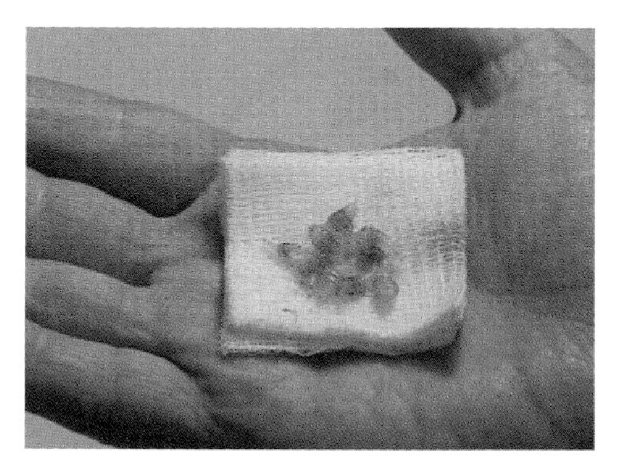

Dr. Keystone holds out the evicted botflies.

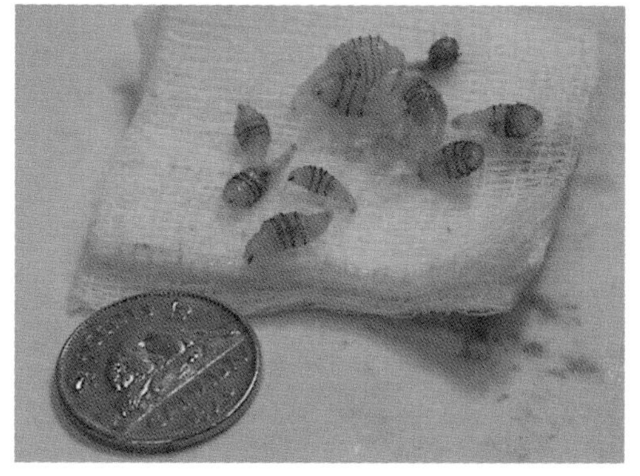

Eleven were removed while one stubborn larva would not be budged and was left to be broken down by the body of its brave host. . .if it didn't pop out on its own first.

One World

We must always remember that we humans have wildlife in us.

Planet Human as a Micro-Environment of Planet Earth

When we come to put it all together, what have we got? The answer is both very simple and very complex. We humans share the occupancy of our planet—as we share the occupancy of our skin—with a vast number of other species. The animals and plants that we see around us and the life forms that we have inside us and on us—in other words the biosphere, or living community of earth—are not separated from us by a distinct and rigid frontier. There is an awful lot of traffic across the border—most of it does no harm, some of it is beneficial to life on both sides, and some of it is distinctly dangerous and destructive. We need to be able to recognize which is which.

This is a mosquito that has just bit into human skin.

And that is perhaps the true message of *Human Wildlife:* it's not some alarmist head-for-the-hills "know the enemy" war cry. It's really an encouraging—even reassuring—call to learn more about the other species, to understand what they can do to us and what they cannot, and then to make our interactions with the rest of life more intelligent and more informed. As has been said many times, the more you know, the less you fear.

Knowledge—Understanding—Better-Informed Interactions

Curiously enough, these same three ingredients also make a good recipe for better co-existence on this planet in every respect—both with the species who are our neighbours and co-tenants, and also among the members of our own species. There are perhaps—and I hope this doesn't sound grandiose or preachy—strong similarities between our reactions to other species and our responses to other cultures. When we are perplexed by an unexpected and unusual form of behaviour, we tend to become defensive, anxious, and, sometimes, hostile. The same is true when we see, for example, strange species for the first time: our initial reactions often involve withdrawal and/or aggression.

A good illustration of this is the way we respond to spiders. In some countries, there are toxic and lethal spiders, but in many places, there is no genuine threat. Yet even where spiders are benign, a proportion of people are genuinely phobic (probably less than one in ten), while many more exhibit revulsion with some measure of fear. Why? Even children recoil the first time they see a spider move in some apparently random and unpredictable way . It doesn't matter whether the parents tell them the spider is not dangerous, they will recoil. To some extent we are programmed—hard-wired—to back away from things we don't understand, and in the course of evolution this was undoubtedly a safe and helpful strategy for survival. Our Neanderthal predecessors wouldn't have fared well if they had attempted to domesticate rhinoceros or tigers or scorpions.

So the first step is to know more—to know which life forms are dangerous and are to be avoided, and which ones are not, and to understand what is likely to happen when we and they meet. Only by that knowledge and that understanding, can we improve the value and usefulness of our interactions.

True as this is with other species, so it is with other cultures within our species. Our reactions to any community that has customs that are different from ours, is often full of mistrust, backed up by anxiety, aggression, and often hostility. Misinterpretation is common because some signals mean different things in different communities. Anyone who has experienced it, can remember the surprise they felt when they first saw someone raise their head backwards and make a clucking sound with their tongue to indicate no, instead of the expected side-to-side shaking of the head. It is a difficult gesture to understand at first, and one's temptation (usually) is to ignore it and try to move forward—almost immediately incurring the wrath of the naysayer. But my whole point is this: once you know that this particular gesture is the common way to signify no, then you can respond appropriately.

Let's face it, as a species we have some pretty impressive abilities. We are capable of vast acts of creative collaboration—building cities, cultivating land, trading, and creating art—but we are also capable of acts of widespread destruction (World War I is an horrific example). Our ability to interact intelligently with members of our own species is as vital as our ability to interact intelligently with other species—in both cases, first and foremost, we need information and understanding.

When all is said and done, life on this planet is extraordinary and constantly surprising, and we are part of that life, not apart from it. With a bit more knowledge and a bit more understanding of how things work, we can use our constructive abilities to greater advantage, and—we all hope—restrain our powerful destructive tendencies more often. As with the biosphere of this planet, as with the ecosystem of our own bodies, so with our relationships with each other—we humans have wildlife inside us in more ways than one.

Acknowledgements

I owe a considerable debt of gratitude to a very considerable number of very considerate people. To begin with, David Langer, the producer of the TV series, and I had many exciting and stimulating conversations with Dr. Kevin Kain (University Health Network, Toronto) and with Dr. Andy Simor (Sunnybrook and Women's Health Sciences Centre, Toronto). We also had a lot of assistance and prompting on a whole range of topics from Dr. Donald Low (Mount Sinai Hospital, Toronto).

The BioMicrotech team of Tim Richardson, Bruno Chue, and Nu-Ann Pham are also unique and talented people whose help was invaluable.

I'd also like to thank the following doctors and associates and their teams for giving illuminating and fascinating interviews:

Michael and Suzanne Sukhdeo, Eric Hoberg, Joel Weinstock, David Elliot, Steve Thomas, Michael Levitt, Carol Storm, Alan Hirsch, Jon Richter, George Preti, Page Caufield, Philip Tierno, Richard Ellen, John McComb, Julian Nedzelski, Neil Shear, Laurence Cohen, Michael Fung Han, Terry Galloway, Craig Stephen, David Kerr, Gregor Reid, Michael Easterbrook, Ernesto Ruiz, Rick Ostfeld, Barbara Yaffe, Tom Baker, Dean Cliver, Anthony Guiterrez, Lia McCabe, and also Jacqueline Duffin and Anne Hardy.

On the television series, I want to mention the incredible collaboration I experienced from David Langer, Elliot Shiff, John Darroch, Liam O'Rinn, Roxanna Spicer, Tim Wolochatiuk, and the incredible camerawork of Mike Ellis, ably assisted by soundman John Martin. Lazlo Barna (Barna-Alpa Productions) and Paul Lewis (Discovery Channel Canada) have given incredibly support to this whole project from the beginning.

Above all this book would not have happened without the indefatigable photo research (and a hundred other activities) of my assistant, Andrew Ferns.

(Opposite)

"These Things, Too, Shall Pass." But is it "shall" or "may"? We all have the feeling that the predominance of the human race is inviolable and will be perpetual. But, as our survey of wildlife has shown, the balance is actually quite fragile. If we don't pay attention to what is going on around us, our future might be no more permanent than footsteps in the sand.

Photographs and Illustrations

Photographs

Page ii: *Dermacentor variabilis* (American dog tick), electron micrograph by Dennis Kunkel, copyright Dennis Kunkel Microscopy, Inc.

Page iv: Top left, *Phthirus pubis*, electron micrograph by Dennis Kunkel, copyright Dennis Kunkel Microscopy, Inc. Bottom right, *Pediculus humanus capitis*, electron micrograph by Dennis Kunkel, copyright Dennis Kunkel.

Page v: Top left, *Cimex Lectularius*, electron micrograph by Dennis Kunkel, copyright Dennis Kunkel Microscopy, Inc. Bottom left, female *Aedes aegypti* mosquito and its proboscis, electron micrograph by Dennis Kunkel, copyright Dennis Kunkel Microscopy, Inc.

Chapter 1

Page 2: *E. coli* on a human hair follicle, electron micrograph by Dennis Kunkel, copyright Dennis Kunkel Microscopy, Inc.

Page 4: Flamingos, photograph by Kevin Kain.

Page 9: Top left; bacteria on a pinhead x 400, electron micrograph courtesy of Nanoworld Images. Top left middle, bacteria on a pinhead x 500, electron micrograph courtesy of Nanoworld Images. Top right middle, bacteria on a pinhead x 1000, electron micrograph courtesy of Nanoworld Images. Top right, *Giardia lamblia*, photograph by Steven J. Upton, Kansas State University. Right middle; *Giardia lamblia*, photograph by Steven J. Upton, Kansas State University. Bottom, pinhead and eye of a needle, electron micrograph by Dennis Kunkel, copyright Dennis Kunkel Microscopy, Inc.

Chapter 2

Page 10: *Pediculus humanus capitis* (head louse), electron micrograph by Dennis Kunkel, copyright Dennis Kunkel Microscopy, Inc.

Page 11: Bottom left, *Demodex foliculorum*, footage by Biomicrotech. Bottom right, *Dermataphagoides farinae* (dust mite) footage by Biomicrotech.

Page12: *Demodex foliculorum* (eyebrow mite) electron micrograph courtesy of Microscopix.

Page 13: *Demodex foliculorum*, footage by Biomicrotech.

Page 14: *Demodex foliculorum*, mites in a follicle, electron micrograph courtesy of Microscopix.

Page 15: *Dermatophagoides farinae* (dust mite), footage by Biomicrotech.

Page 16: Top right and left, *Dermatophagoides farinae* (dust mite with anal pellet), footage by Biomicrotech.

Page 18: *Sarcoptes scabei* (scabies mite), electron micrograph by Louis De Vos.

Page 19: *Ixodes scapularis* (black-legged deer tick), footage by Michael Ellis.

Page 20: Top, *Ixodes scapularis*, electron micrograph Dennis Kunkel, copyright Dennis Kunkel Microscopy, Inc. Bottom left and right, *Ixodes scapularis*, male and female, electron micrograph by Louis De Vos.

Page 21: Top,Dr. Rick Ostfeld, footage by Michael Ellis. Bottom, mouse, footage by Michael Ellis. Page 22: Top left, tick on finger, footage by Michael Ellis. Top right, *Ixodes scapularis* eggs, footage by Michael Ellis. Middle left, *Ixodes scapularis* nymph and an engorged female, footage by Michael Ellis. Middle right; *Ixodes scapularis*, footage by Michael Ellis. Bottom left, engorged *Ixodes scapularis* females and a nymph, footage by Michael Ellis.

Page 23: Mouth parts of the *Dermacentor variabilis* (American dog tick), electron micrograph by Dennis Kunkel, copyright Dennis Kunkel Microscopy, Inc.

Page 24: Top left, middle and right, Lia McCabe in the woods, footage by Michael Ellis.

Page 25: Underbelly of *Ixodes scapularis*, footage by Michael Ellis.

Page 26: Chiggers on human skin, electron micrograph by Dennis Kunkel, copyright Dennis Kunkel Microscopy, Inc.

Page 27: Removed chigger mite with eggs, courtesy of Kevin Kain.

Page 28: Top, chigger bite wound, courtesy of Kevin Kain. Bottom, the author and "Buddy," photograph by David Langer.

Page 29: Left, *Ctenocephalides felis*, electron micrograph by Dennis Kunkel, copyright Dennis Kunkel Microscopy, Inc. Top right, *Ctenocephalides felis* (cat flea), footage by Biomicrotech. Middle right, leg of a *Ctenocephalides felis*, footage by Biomicrotech. Bottom right, flea eggs, footage by Biomicrotech.

Page 30: Top left, Dr. Terry Galloway, entomologist, footage by Michael Ellis. Middle, *Pulex irritans*, electron micrograph by Louis De Vos.

Page 32: *Cimex Lectularius*, electron micrograph by Dennis Kunkel, copyright Dennis Kunkel Microscopy, Inc.

Page 33: Left, *Cimex Lectularius*, footage by Michael Ellis. Right, *Cimex Lectularius* (bedbug) footage by Biomicrotech.

Page 34: Top left, *Pediculus humanus capitis* (head louse), footage by Biomicrotech. Top right, nits, footage by Biomicrotech. Middle left, Dawn Muchie, professional nit-picker of the lice squad, footage by Michael Ellis. Middle right, a nit, footage by Biomicrotech. Bottom right, *Pediculus humanus capitis*, footage by Biomicrotech.

Page 35: Top, claw of *Pediculus humanus capitis*, electron micrograph courtesy of Nanoworld Images. Bottom, nit of the *Pediculus humanus capitis*, electron micrograph by Louis De Vos.

Page 36: *Phthirus pubis*, electron micrograph by Dennis Kunkel, copyright Dennis Kunkel Microscopy, Inc.

Page 37: *Phthirus pubis* (crab louse), in an eyelash, photograph by Dr. Michael Easterbrook.

Chapter 3

Page 38: Left, middle, right, photographs by Kevin Kain.

Page 41: Top, pond scum, footage by Michael Ellis. Middle, bacteria colonies in a biofilm, footage by Biomicrotech. Bottom, bacteria colonies in a biofilm, footage by Biomicrotech.

Page 42: Plaque and tartar on human tooth enamel, electron micrograph by Dennis Kunkel, copyright Dennis Kunkel Microscopy, Inc.

Page 43: Left, the human mouth, footage by Michael Ellis. Middle, saliva inside the human mouth, footage by Michael Ellis. Right, bacteria colonies in a biofilm, footage by Biomicrotech.

Page 45: Top, Spirochete bacteria in the human mouth, footage by Biomicrotech. Bottom, man and his tongue, footage by Michael Ellis.

Page 46: Top left, Dr. Michael Richter, footage by Michael Ellis. Top right, Dr. Richter and patient, footage by Michael Ellis. Bottom left, swabbing the tongue, footage by Michael Ellis. Bottom right, tongue bacteria, footage by Biomicrotech.

Page 47: Top, middle, bottom, vial with tongue bacteria, footage by Michael Ellis.

Page 48: Left, Toronto, footage by Michael Ellis. Right, New York City, footage by Michael Ellis.

Page 49: Left, pedestrians in New York, footage by Michael Ellis. Right, Four Seasons Hotel, Toronto, footage by Michael Ellis.

Page 50: Left, *Candida albicans* on a human tongue, footage by Michael Ellis. Right, *Candida albicans*, electron micrograph by Dennis Kunkel, copyright Dennis Kunkel Microscopy, Inc.

Page 52: Top left, Dr. Page Caufield, New York University, footage by Michael Ellis. Top right, bacteria of the human mouth, footage by Biomicrotech. Bottom, breast-feeding baby, footage by Michael Ellis.

Page 54: Top left: Contact lens, footage by Michael Ellis. Top right, human tongue and finger, footage by Michael Ellis. Bottom left, contact lens and eye, footage by Michael Ellis. Bottom right, human eye, footage by Michael Ellis.

Page 55: Top left, Dr. Andy Simor, Sunnybrook and Women's College Medical Sciences Centre, footage by Michael Ellis. Top middle, bacterial culture on an Agar plate, footage by Michael Ellis. Top right, bacterial culture on an Agar plate with the word "tears'" written with tears prior to culturing, footage by Michael Ellis. Bottom, the author outside the Monell Chemical Senses Center, Philadelphia, photograph by Michael Ellis.

Chapter 4

Page 56: Scolex of *Hymenolopsis diminunata* (rat tapeworm), electron micrograph by Mike Kiser, University of Virginia at Wise, courtesy of Prof. J. Rex Baird.

Page 57: Eggs of the *Hymenolopsis diminunata* (rat tapeworm), electron micrograph by Mike Kiser, University of Virginia at Wise, courtesy of Prof. J. Rex Baird.

Page 59: Top, Dr. Michael Sukdheo, Rutgers University. Bottom left, *Trichinella spiralis* cysts, photograph by Steven J. Upton, Kansas State University. Bottom middle, *Trichinella spiralis* cysts, photograph by Steven J. Upton, Kansas State University. Bottom right, a free swimming *Trichinella spiralis*, photograph by Steven J. Upton, Kansas State University.

Page 61: Top; *Ascaris lumbricoides*, electron micrograph courtesy of Nanoworld Images bottom; *Ascaris* emerging from pig intestine, footage by Michael Ellis.

Page 62: Top, *Ascaris* on human hand, footage by Michael Ellis. Middle, *Ascarid* egg, photograph by Steven J. Upton, Kansas State University. Bottom, *Ascaris* larvae inside egg, footage by Biomicrotech.

Page 63: Top left, *Fasciolopsis buski*, giant intestinal fluke, photograph by Steven J. Upton, Kansas State University. Top right, *Clonorchis sinensis*, Chinese liver fluke, photograph by Steven J. Upton, Kansas State University. Bottom, *Fasciolopsis buski*, giant intestinal fluke, photograph by Steven J. Upton, Kansas State University.

Page 64: Left, Dr. J.D. MacLean, Centre for Tropical Diseases, McGill University, footage by Michael Ellis. Top and bottom right, *Metorchis conjunctis* (North American liver fluke), footage by Michael Ellis.

Page 65: Left, nonclassified fluke, electron micrograph by Louis De Vos. Middle, *Shistosoma* egg, photograph by Steven J. Upton, Kansas State University. Right, Shistosomes "in copula," female inside the male, photograph by Steven J. Upton, Kansas State University.

Page 66: Boy with snails, China, photograph by Kevin Kain.

Page 67: Mauritius, photograph by Kevin Kain.

Page 68: Top: *Ancylostoma duodenale* (hookworm eggs), photograph by Steven J. Upton, Kansas State University. Bottom, *Ancylostoma duodenale* (hookworm), electron micrograph, courtesy of Nanoworld Images.

Page 69: Top left, *Heligmosomoides polygyrus*, footage by Michael Ellis. Top right, *Heligmosomoides polygyrus* larvae on charcoal, footage by Michael Ellis. Bottom, Cutaneous larval migrans, photograph courtesy of Kevin Kain.

Page 70: Top; Proglottids of the *Taenia saginata* (beef tapeworm), photograph by Steven J. Upton, Kansas State University. Bottom left, Scolex of the *Taenia pisiformis*, photograph by Steven J. Upton, Kansas State University. Bottom right, Scolex of the *Taenia pisiformis*, photograph by Steven J. Upton, Kansas State University.

Page 72: Top, copepod, footage by Biomicrotech. Bottom, copepod, footage by Biomicrotech.

Page 73: Rope bridge in New Guinea, photograph by Geoff Ibbotson.

Page 74: Top left, *Dracunculus mediensis*, guinea worm being extracted from a young boy's foot, footage courtesy of Foster Parents Plan. Top right, *Dracunculus mediensis* extraction, footage courtesy of Foster Parents Plan. Bottom, *Dracunculus mediensis* emerged from foot, photograph courtesy of Kevin Kain.

Page 75: Left, right, *Dracunculus mediensis* extraction from foot, footage courtesy of Foster Parents Plan.

Page 76: Top, *Toxocara catis* (*Toxocara* worm), electron micrograph, courtesy of Nanoworld Images. Bottom left, *Toxocara catis* egg, photograph by Steven J. Upton, Kansas State University. Bottom right, *Toxocara catis* worm in a human eye, photograph by Dr. Michael Easterbrook.

Page 77: Top left, *Toxocara* larvae exiting its egg, footage by Biomicrotech. Top right, *Toxocara catis* larvae and its egg, footage by Biomicrotech. Bottom left, *Toxocara catis* eggs and larvae, footage by Biomicrotech. Bottom right, *Toxocara catis* larvae, footage by Biomicrotech.

Page 79: Top left, right, bottom, *Lucilia sericata*, greenbottle fly on liver, footage by Michael Ellis.

Page 80: *Lucilia sericata* larvae, medicinal maggots on a finger, photograph courtesy of Dr. Steven Thomas.

Page 81: Top left, *Lucilia sericata* larvae, footage by Michael Ellis. Top right, Dr. Steven Thomas, Surgical Materials Testing Laboratory, Princess of Wales Hospital, Bridgend, UK, footage by Michael Ellis. Bottom left, *Lucilia sericata* larvae in a wound, footage by Michael Ellis. Bottom right, *Lucilia sericata* larvae on a wound, footage by Michael Ellis.

Page 82: Left, *Lucilia sericata* larvae on a needle, photography courtesy of Dr. Steven Thomas. Right, application of larvae and bandages to a wound, footage by Michael Ellis.

Page 84: Left, *Trichuris* egg, footage by Biomicrotech. Right, preparation of "Worm Cocktail," footage by Michael Ellis.

Page 85: Left, Dr. Joel Weinstock, University of Iowa, footage by Michael Ellis. Right, Dr. David Elliot, University of Iowa, footage by Michael Ellis.

Page 86: Left, *Trichuris* larvae, footage by Biomicrotech, right. *Trichuris* egg with larva inside, footage by Biomicrotech.

Page 88: The author at the University of Iowa, footage by Michael Ellis.

Chapter 5

Page 90: The human armpit, photograph by Brooke Palmer.

Page 92: Amazon jungle, photograph by Kevin Kain.

Page 93: Left, armpit bacteria sampling, footage by Michael Ellis. Middle, armpit swabbing, footage by Michael Ellis. Right, armpit bacteria, footage by Biomicrotech.

Page 94: Dr. George Preti taking a sample, footage by Michael Ellis.

Page 95: Women sniffing T-shirts, footage by Michael Ellis.

Page 96: Nose in a smell test machine, footage by Michael Ellis.

Page 97: Kissing couple, footage by Michael Ellis.

Page 98: Dr. Alan Hirsch, Smell and Taste Treatment Center in Chicago, footage by Michael Ellis.

Page 101: *Propionobacteria acnes*, electron micrograph by Dennis Kunkel, copyright Dennis Kunkel Microscopy, Inc.

Page 102: *Propionobacteria*, footage by Biomicrotech.

Page 103: Acne, photograph courtesy of Dr. Neil Shear, Department of Dermatology, Sunnybrook and Women's College Health Sciences Centre.

Page 104: *Epidermophyton floccosum*, an Athlete's Foot fungi, electron micrograph by Dennis Kunkel, copyright Dennis Kunkel Microscopy, Inc.

Page 105: Left, Athlete's Foot, photograph courtesy of Dr. Neil Shear, Department of Dermatology, Sunnybrook and Women's College Health Sciences Centre. Top right, Athlete's Foot fungi, footage by Biomicrotech. Middle right, Jock Itch fungi, footage by Biomicrotech. Bottom right, Jock Itch fungi, footage by Biomicrotech.

Page 107: Top, *Herpes simplex virus* (HSV6) on a blood lymphocyte, electron micrograph by Dennis Kunkel, copyright

Dennis Kunkel Microscopy, Inc. Bottom left, wart on a thumb, footage by Michael Ellis. Bottom middle, wart on a thumb, footage by Michael Ellis. Bottom right, laser treatment of a wart on a foot, footage by Michael Ellis.

Page 109: *Toxoplasma gondii*, electron micrograph by Dennis Kunkel, copyright Dennis Kunkel Microscopy, Inc.

Page 110: Left, cat in a litter box, footage by Michael Ellis. Right, Dr. Craig Steven, footage by Michael Ellis.

Page 111: *Toxoplasma gondii* oocyst, photograph by Steven J. Upton, Kansas State University.

Page 112: Top left, Carol Storm getting into her fart pants, footage by Michael Ellis. Top right, Dr. Michael Levitt, footage by Michael Ellis. Bottom, bacteria of the colon, footage by Biomicrotech.

Page 114: Left, the fart pants, footage by Michael Ellis. Right, extraction of gases from fart pants, footage by Michael Ellis.

Page 115: Bacteria of the colon, footage by Biomicrotech.

Page 116: Smelling the gases, footage by Michael Ellis.

Chapter 6

Page 118: An Amazon waterfall, photograph by Kevin Kain.

Page 123: Top left, Dr. Charles Gerba toilet sample, footage by Michael Ellis. Top middle, Dr. Charles Gerba toilet sample, footage by Michael Ellis. Top right, Dr. Charles Gerba toilet sample, footage by Michael Ellis. Bottom left, Dr. Gerba's luminometer, footage by Michael Ellis. Bottom right, culture of *E. coli* from samples, footage by Michael Ellis.

Page 124: Top left, Dr. Charles Gerba and washing machine, footage by Michael Ellis. Top right, Dr. Charles Gerba and samples, footage by Michael Ellis. Bottom, Dr. Charles Gerba and kitchen, footage by Michael Ellis.

Page 125: *E. coli OH157*, electron micrograph by Dennis Kunkel, copyright Dennis Kunkel Microscopy, Inc.

Page 126: Kitchen sponge and its inhabitants, electron micrograph by Dennis Kunkel, copyright Dennis Kunkel Microscopy, Inc.

Page 127: Plastic cutting board and inhabitants, electron micrograph by Dennis Kunkel, copyright Dennis Kunkel Microscopy, Inc.

Page 131: The author at the Broadstreet Pump, London, UK, footage by Michael Ellis.

Page 132: Left, Anne Hardy, footage by Michael Ellis.

Page 133: *Vibrio cholerae*, electron micrograph by Dennis Kunkel, copyright Dennis Kunkel Microscopy, Inc.

Page 134: A Hong Kong market street, photograph by Kevin Kain.

Page 135: Top, New Guinea, photograph by Kevin Kain. Inset, sewage treatment plant, Toronto, footage by Michael Ellis.

Page 137: Left, right, *Cyclospora*, photograph by Steven J. Upton, Kansas State University.

Page 138: Raspberry, footage by Michael Ellis.

Page 139: Top left, raspberries, footage by Michael Ellis Top right, *Cyclospora*, footage by Biomicrotech. Bottom, Dr. Barbara Yaffe, footage by Michael Ellis.

Page 140: The author at a fruit market, footage by Michael Ellis.

Page 141: Top, Dr. Dean Cliver in Milwaukee, footage by Michael Ellis. Bottom left, *Cryptosporidium* infection of the intestine, electron micrograph by W.L. Current, courtesy of Steven J. Upton, Kansas State University. Bottom middle, *Cryptosporidium*, electron micrograph by W.L. Current, courtesy of Steven J. Upton, Kansas State University. Bottom right, *Cryptosporidium*, electron micrograph by W.L. Current, courtesy of Steven J. Upton, Kansas State University.

Page 142: *Cryptosporidium*, electron micrograph by W.L. Current, courtesy of Steven J. Upton, Kansas State University.

Page 143: Left, *Cryptosporidium*, electron micrograph by W.L.

Current, courtesy of Steven J. Upton, Kansas State University. Right, *Cryptosporidium*, electron micrograph by W.L. Current, courtesy of Steven J. Upton, Kansas State University.

Page 144: Alberta, photograph by Kevin Kain.

Page 145: Top, *Giardia lamblia*, photograph by Steven J. Upton, Kansas State University. Bottom, a beaver's lodge, footage by Michael Ellis.

Page 146: *Giardia* infection of the intestine, electron micrograph by Dennis Kunkel, copyright Dennis Kunkel Microscopy, Inc.

Page 147: *Salmonella typhi*, electron micrograph by Dennis Kunkel, copyright Dennis Kunkel Microscopy, Inc.

Page 148: *Campylobacter jejuni*, electron micrograph by Dennis Kunkel, copyright Dennis Kunkel Microscopy, Inc.

Page 150: Top, the author and crew at a chicken processing plant, Canada, photograph by Andrew Ferns. Middle, turning the chickens insides out, photograph by Andrew Ferns. Bottom, individual inspection of chickens and their insides, photograph by Andrew Ferns.

Page 152: Top, *Lactobacilli*, footage by Biomicrotech. Bottom, *Lactobacilli,* footage by Biomicrotech.

Page 153: Top; *Lactobacilli* descending on vaginal wall, computer-generated images by Mark Neysmith. Bottom left, *Lactobacilli* and *E. coli* on vaginal wall, computer-generated images by Mark Neysmith. Bottom right, Dr. Gregor Reid, footage by Michael Ellis.

Page 154: *Enterococci faecium,* electron micrograph by Dennis Kunkel, copyright Dennis Kunkel Microscopy, Inc.

Page 155: *Trichomonas vaginalis*, protozoan parasite, electron micrograph by Dennis Kunkel, copyright Dennis Kunkel Microscopy, Inc.

Page 157: Left, Dr. Phillip Tierno, footage by Michael Ellis. Right, Dr. Page Caufield, footage by Michael Ellis.

Page 158: Hand washing, footage by Michael Ellis.

Chapter 7

Page 160: Main, truck near the Khyber Pass, Afghanistan, photograph by Kevin Kain. Top left, *Salmonella typhi*, footage by Biomicrotech. Top right, chicken, photograph by David Langer. Middle, sheep and chicken, photograph by David Langer. Bottom left, *Plasmodium vivax*, malarial parasite, photograph by Steven J. Upton, Kansas State University. Bottom middle, mosquito taking a blood meal, footage courtesy of Barna Alper Productions. Bottom right, the truck's driver, photograph by Kevin Kain.

Page 161: *Ctenocephalides felis*, footage by Michael Ellis.

Page 162: Author and leaf cutter ants, Costa Rica, footage by Michael Ellis.

Page 164: Pagang, Burma, photograph by Kevin Kain. Inset, *E. coli*, electron micrograph by Dennis Kunkel, copyright Dennis Kunkel Microscopy, Inc.

Page 165: Rice paddies, South East Asia, photograph by Kevin Kain.

Page 166: Left, female *Aedes aegypti* mosquito and its proboscis, electron micrograph by Dennis Kunkel, copyright Dennis Kunkel Microscopy, Inc. Right, *Plasmodium vivax* sporozoites in a mosquito's saliva, photograph by Steven J. Upton, Kansas State University.

Page 167: Left, Asian tiger mosquito preparing to take a blood meal, electron micrograph by Dennis Kunkel, copyright Dennis Kunkel Microscopy, Inc. Right, *Plasmodium vivax* oocysts on a mosquito's gut wall, photograph by Steven J. Upton, Kansas State University.

Page 169: *Plasmodium falciparum* macrogametocyte with red blood cells, electron micrograph by Oliver Meckes/Nicole Ottawa/Photo Researchers, Inc.

Page 170: Top left, *Plasmodium falciparum* ring stages in blood. Top right, *P. falciparum* macrogametocytes in blood. Middle left, *P. vivax* ring stage in a blood cell, *P. vivax* meront in a blood cell. Middle right, *P. vivax* meronts in blood cells.

Bottom left, *P. vivax* gametocyte in a blood cell. All photographs by Steven J. Upton, Kansas State University.

Page 172: Top left, Dr. Anthony Guiterez, US Army Center for Health Promotion and Preventative Medicine, Aberdeen Proving Ground, Maryland, footage by Michael Ellis. Top right, heads-up display of new device at CHPPM, footage by Michael Ellis. Bottom, Dr. Guiterez and portable lab at CHPPM, footage by Michael Ellis.

Page 174: Top left, *Trypanosoma gambiense*, trypomastigoites in blood, photograph by Steven J. Upton, Kansas State University. Right, *Trypanosoma cruzi*, trypomastigoites in blood, electron micrograph by Oliver Meckes/Nicole Ottawa/Photo Researchers, Inc. Bottom left, *Trypanosoma gambiense*, trypomastigoites in blood, photograph by Steven J. Upton, Kansas State University.

Page 176: *Legionella pneumophilia*, electron micrograph by Dennis Kunkel, copyright Dennis Kunkel Microscopy, Inc.

Page 178: *Helicobacter pylori*, electron micrograph by Dennis Kunkel, copyright Dennis Kunkel Microscopy, Inc.

Page 183: Bacon on a botfly wound, photograph by Jay Keystone, courtesy of Kevin Kain.

Page 184: Left, botfly emerged from wound after bacon treatment, photograph by Jay Keystone, courtesy of Kevin Kain. Top right, bottom right, *Lucilia sericata*, greenbottle fly, footage by Michael Ellis.

Page 185: Botfly larvae in arm, photograph by Jay Keystone, courtesy of Kevin Kain.

Page 186: Top left, removal of bandage. Top right, using lateral pressure to remove botfly larvae. Middle left, botfly emerging. Middle right, another botfly Eviction. Bottom, extracted botfly larvae next to wound, all photographs by Andrew Ferns.

Page 187: Top, the botfly collection so far. Middle, evicted larvae well in hand. Bottom, all the removed botflies. All photographs by Andrew Ferns.

Chapter 8

Page 188: Rowing on the Mekong, photograph by Kevin Kain.

Page 190: Mosquito taking a blood meal, footage courtesy of Barna Alper Productions.

Page 192: Namibian desert, photograph by Kevin Kain.

Illustrations

Index

Page citations in **boldface** below refer to photographs in the book.